GED® MATH
TEST TUTOR

$$30\% = .30 = \frac{3}{10}$$

Sandra Rush
B.A. (Temple University)
M.A. (University of California, Los Angeles)

Research & Education Association
Visit our website: www.rea.com

Research & Education Association
61 Ethel Road West
Piscataway, New Jersey 08854
E-mail: info@rea.com

GED® Math Test Tutor

Printed in the United States of America

Library of Congress Control Number 2013914515

ISBN-13: 978-0-7386-1136-5
ISBN-10: 0-7386-1136-0

REA® is a registered trademark of Research & Education Association, Inc.

Contents

Chapter 1: About the Next-Generation GED® Math Test

Chapter 2: "I Was Born Knowing That"

Chapter 3: The Parts of the Whole

Chapter 4: Power Play

Chapter 5: Algebra—Not a Four-letter Word

Chapter 6: The X (Squared) Factor

Chapter 7: The Shape of Things

Chapter 8: What Are the Chances?

GED® Math Practice Test 1   249

GED® Math Practice Test 2   269

Glossary   289

Index   297

About the Author

Sandra Rush earned a Bachelor of Arts degree in mathematics from Temple University in Philadelphia. While an undergraduate student, she served as a math tutor for members of the basketball team as well as for local public school students. In these early years of her own higher education, she realized that teaching mathematics had a special appeal.

When awarded a research assistantship and later a Ford Foundation Fellowship for an advanced degree in the Department of Atmospheric Sciences at the University of California, Los Angeles, Sandra pursued graduate school and earned a Master of Arts degree in ionospheric physics. Upon graduation, however, she turned down offers to become a physicist in favor of a career as a mathematics instructor at the secondary school level. She quickly became chair of the mathematics department at Notre Dame Academy in Massachusetts.

At about that same time, she tutored the first of several young adults pursuing a high school equivalency certificate. That student passed the math portion of the test on her first attempt, and with the certificate she was able to advance in her career. Sandra was "hooked on teaching." She has taught mathematics and physics from the junior high school to university levels in three states. Sandra has also continued her side career of tutoring and coaching students at all levels. Her one-on-one approach with young adults has yielded excellent results.

Sandra's interest in education has extended beyond the classroom to the field of publishing, including writing and editing manuals on preparation for standardized tests at all levels. This book is the latest of her efforts to make mathematics both understandable and enjoyable for generations of students.

Acknowledgments

I think that people read acknowledgments just to see whether their names are mentioned (and spelled correctly). Authors write acknowledgments to give thanks to people who have been significant in encouraging their efforts. My list begins with my children, Mike and Kara, who have always supported me in my endeavors, no matter how bizarre they sometimes seemed. My twin sister Sue and friends Joann, Carol, Dave, Virgil, and Herb have provided much-appreciated day-to-day support and love. My longtime friends from Humanities have influenced me and given me confidence in my abilities. Thank you all, even those who touched my life but whose names aren't specifically mentioned here.

Author Dedication

This book is dedicated in memory of two of my high school teachers,

Erasmo S. Ciccolella,

who taught me how to write, and

Frederick W. Drewes,

who encouraged my love of math.

What a profound and lasting influence great teachers have on their students.

About Correlations to GED® Assessment Targets

We would like to thank Creative Services Associates (CSA) for conducting a detailed independent review of this publication. CSA confirms that all of the GED® 2014 Mathematical Reasoning and Mathematical Practices Assessment Targets are covered through both the review materials and the practice tests. CSA recommends this text to test-takers seeking the tools to be successful with the mathematical portion of the new version of the GED® test.

About Research & Education Association

Founded in 1959, Research & Education Association (REA) is dedicated to publishing the finest and most effective educational materials—including study guides and test preps—for students in middle school, high school, college, graduate school, and beyond.

Today, REA's wide-ranging catalog is a leading resource for teachers, students, and professionals. Visit *www.rea.com* to see a complete listing of all our titles.

REA Acknowledgments

We would like to thank Leanne Fucci for her technical edit. Ms. Fucci has a B.S. in Education and an M.S. in Educational Administration. She has taught Adult Basic Education and GED® Instruction for 17 years—for the State of New Jersey in the New Jersey Department of Corrections to female inmates, ranging in age from 16 to 63 years old. We extend gratitude to GED® math test author and tutor Steve Reiss for his readthrough and perspective. We would also like to thank Pam Weston, Publisher, for setting the quality standards for production integrity and managing the publication to completion; John Paul Cording, Vice President, Technology, for coordinating the design and development of the REA Study Center; Larry B. Kling, Vice President, Editorial, for his overall direction; Michael Reynolds, Managing Editor, for coordinating development of this edition; Transcend Creative Services for typesetting this edition; Claudia Petrilli, Graphic Designer, for designing our interior; and Weymouth Design and Christine Saul, Senior Graphic Designer, for designing our cover.

REA gratefully acknowledges Texas Instruments' permission to use an image of their TI-30XS Multiview calculator.

Chapter 1

About the Next-Generation GED® Math Test

This introductory chapter is as important as any of the following chapters. The subject review chapters will show you what you need to know about mathematics for the GED® test, but if you don't know *how* to take the test and what to expect, you may not be successful, even if you understand the math. To be successful with the GED® Mathematical Reasoning test, it is essential that you know three skill sets: math skills, calculator skills, and some computer skills.

How This Book Can Help

For all three skill sets, practice, practice, practice will pay off. This book helps with the mathematical reasoning skills, so read the book carefully and be sure you understand each section before you advance to the next. Numerous examples will give you the practice and confidence you need to do well. To paraphrase the GED® Testing Service, it's not just about knowing how, but rather why!

Remembering isn't the same as memorizing. This book is designed to help you remember facts that you need without having to memorize. Each mathematical topic is introduced, accompanied by several example questions with answers and explanations that are designed as further teaching aids to help you remember what has been presented. Exercises at the end of each chapter allow you to assess how well you understood the concepts introduced in the chapter. These exercises are presented in GED® test formats so you can get used to how questions will be presented on the actual test. Answers and explanations for those answers appear after the exercises.

You probably have heard the saying "We learn from our mistakes." If you do make a mistake on any of the exercises, use that as a further learning experience and don't be discouraged. If you ever locked the key in your car, that was a learning experience (a frustrating one, for sure) and, at least for a while after that, you always made sure you had the key in your hand when you locked the door. So if you choose the wrong answer to an exercise in the book, go to the explanation of the correct answer and learn from it so you won't make that type of mistake again.

The last half of this chapter presents step-by-step instructions for using the TI-30XS virtual calculator that will be provided on the test. In addition, calculator instructions are given in the chapters for which a certain calculation is relevant.

The final part of the book contains two full-length practice GED® Mathematical Reasoning tests followed by the correct answer choices with explanations. Time yourself (90 minutes) to get a sense of how quickly you must come up with answers. You can go back to any question at any time and you can change your answers if you want to, but you need to do it all within the 90-minute time frame. Admittedly, taking the test on paper is different from taking the test on a computer, but the goal here is to learn the math to answer the problems correctly.

As of the printing of this book, the GED® Testing Service is continuing to refine aspects of the GED® Math Reasoning test that will be available in January 2014. Continuing releases of test information from the GED® Testing Service are available on the official GED® website, *www.gedtestingservice.com*.

What to Expect on the GED® Mathematical Reasoning Test

The latest version of the GED® math test is called "Mathematical Reasoning." Previously, the GED® test emphasized mathematical skills, knowing mathematical facts and formulas, and how to perform specific calculations for a problem. Starting in 2014, the emphasis is on understanding and logical thinking rather than memorization, and it is based on real-world problem-solving skills. The next-generation test expects the test-taker to understand what a problem is asking for and proceed to answer it based on the information given. Use the four-part plan for reasoning skills to answer the questions:

1. Figure out what is known, what is necessary to solve the problem, what information is missing (usually your solution), and what is unnecessary (just because something is mentioned doesn't mean it has anything to do with your solution).

2. Devise a strategy to solve the problem. This may involve making a sketch or a table (use the erasable note board) or looking for a pattern, for example.

3. Solve the problem according to your strategy and choose or enter the correct answer.

4. Make sure the answer makes sense.

As an example of the reasoning aspect of the new GED® test, rather than asking "10 − (5 + 2.21) = ?" as might have appeared on previous GED® math tests, the 2014 version asks, "If Liz buys two items at the store that cost $5.00 and $2.21, how much change will she receive from a $10 bill?" It is the same arithmetic, but the test-taker has to know not only how to set up that arithmetic, but also how to answer the question in the computerized format of the test.

The problems on the 2014 GED® test mimic the real world. Almost everyone is carrying around a calculator—it's in the form of a cell phone. Almost no one multiplies 63 × 24 on paper, much less in their heads—they all reach for the calculator. Many store clerks don't count out change anymore, they just give the customer what the register (also a calculator) tells them to. In addition, with so much knowledge available via the Internet, memorization of many formulas is no longer necessary because people can look them up and avoid "misremembering." So the GED® Math Reasoning test measures how well you can figure out what is asked for and then apply mathematical skills, whether from your own memory bank or by using the tools that are available on the test, to come up with answers.

The total test time is 90 minutes, but that includes 2 minutes for an introduction and wrap-up, so the actual testing time is only 88 minutes.

The test is worth 50 points, but there aren't necessarily 50 questions. Some questions are worth twice as much as others—that is because they are twice as difficult. Also, one chart, table, or problem scenario may have several questions associated with it. The computer uses a split screen for this type of question. On a split screen, the graphic or scenario stays on the left side of the screen, and the questions appear one by one on the right screen. Fortunately, there is a "Previous" button if any of these questions depend at all on a response to a prior question.

Fifty-five percent of the questions involve algebra, but that doesn't mean you are just given an equation and asked to solve it. Questions on any of the math subjects on the test (each of which is covered in a chapter of this book) may involve algebraic reasoning. The other 45 percent of the questions are computational.

The scoring on the test is strictly the number of correct answers, with no penalty deducted for wrong answers or bad guesses. Therefore, you should *answer every question*. Once you have selected an answer to a question, take an extra second or two to make sure it makes sense. As obvious examples, the change you receive from a purchase shouldn't be more than the money you gave the clerk, or the discounted cost of an item shouldn't be more than the original price.

Types of Questions

The GED® Math Reasoning test has five types of questions: multiple-choice, drop-down, fill-in-the-blank, hot-spot, and drag-and-drop.

Multiple-Choice Items

The usual **multiple-choice** questions are a major part of the test, but not all of the questions are multiple-choice. These questions have four answers from which to choose, designated A, B, C, and D. You don't necessarily have to come up with the correct answer from scratch—you only have to be able identify the correct answer among the four choices. You should be able to eliminate one or more of the answer choices without any calculation if they are obviously wrong (for example, the problem asks for a whole number and one answer choice is a fraction or decimal), or inappropriate (for example, the problem indicates that the answer will be in the set {1, 2, 3, 4, 5} and the answer choice is 6). To choose an answer, click next to the letter for that choice.

Drop-Down Items

Drop-down items are a variation of multiple-choice questions. These are responses that are embedded directly in the text, and you click on one of three to five given choices to make the statement true. Here's a simplified example: "The number 3 is $\left\{\begin{array}{c}\text{less than}\\\text{equal to}\\\text{greater than}\end{array}\right\}$ the value of 2(5) − 7."

Fill-in-the-Blank Items

Fill-in-the-blank questions involve problems for which you type your answer in a box. It can be the value you got from doing a specific calculation, a one-word or one-phrase answer to a question, or an equation you would use to solve a problem. Fill-in-the-blank questions may have more than one blank to fill in. These questions are similar to multiple-choice questions except that you aren't given any choices for your answer. You must come up with the answer on your own and type it in a box that is part of the question. Don't be concerned about entering extra spaces as you type your answer because the extra spaces are automatically deleted when the test is scored.

A symbol selector sheet is provided on-screen so that you can use symbols in your answer. Each question of this type will state which symbols may be used in your answer. See the section Symbol Selector Sheet in this chapter.

Hot-Spot Items

Hot-spot items require some computer skill in moving the cursor exactly where you want it. These items accompany some sort of graphic on which you click a location to insert a point on a coordinate grid, number line, dot plot, or an edge on a geometric figure. If you change your mind about the placement of a point, clicking that point again will delete it and you will be able to insert a new point on the graph. Occasionally, you will be asked to plot, say, three points. If you try to insert a fourth one, you will be prompted that you have already inserted the requested number of points. You can then delete one of the other points to replace it with the new point.

Drag-and-Drop Items

Likewise, **drag-and-drop** items require computer dexterity in moving the cursor to drag small images on the screen, such as words or parts of numerical expressions (e.g., numbers, operators, or variables), and drop them into boxes where they make sense. Typically, you are given some choices to drag into boxes. For example, you may write an inequality by dragging 5 items from a choice of 8 items: x, $+$, $7y$, $>$, and 9, in that order, to write $x + 7y > 9$. The test format allows you to replace any choice you have made by simply dragging a new item onto where the old item was. Then the old item goes back into the area of choices and the new item you selected goes into the proper box in the question.

Online Resources

The test provides many helpful resources (virtual calculator, calculator reference sheet, a formula sheet, and erasable note boards that are to be used as scratchpads—even a time clock so you can pace yourself). It is essential that you be familiar with each of these so you can easily use them during the test. They will cut down on your test-taking time as well as help with the accuracy of your answers.

The two choices at the lower right corner of each screen are important to bear in mind. One is "Next," which you have to click to go to the next question. The other is "Previous," which comes in handy when all of a sudden you remember how to do a former question, or when you need the results from the last question to answer the present question. This may happen in a split-screen situation, where a graphic or table, for example, is displayed on the left side of the screen, and two or more questions about the same visual are asked on the right side of the screen.

Calculator

The GED® test is completely computer-based. Therefore, just as important as sharpening your mathematical reasoning skills is sharpening your computer and calculator skills. Calculator pages at the end of this chapter give step-by-step instructions for using the virtual TI-30XS calculator that is supplied on the test as a drop-down item on 90% of the questions.

The virtual calculator will help you do the computations, so you won't have to worry about making mathematical errors—the calculator does most of the computations for you. Your main focus should be on understanding what is being asked. For that reason, this book presents the skills needed to do a problem and then lots of "word" problems so you can improve your mathematical reasoning skills. The actual computations are the easier parts of the problems.

Another drop-down item is a calculator reference sheet, which is an abbreviated chart that will give instructions on using the virtual calculator during the test. Even though you will have this sheet to help you during the test, it is highly recommended that you become familiar with a calculator prior to taking the test so you are not spending time during the test learning what you should already know. The step-by-step instructions provided at the end of this chapter are more detailed than the summary of steps provided on the calculator reference sheet.

Formula Sheet

A formula sheet is a drop-down item on the GED® test. It provides essential information necessary for answering problems. This formula sheet provides:

- The areas of a parallelogram and trapezoid

- The surface area and volume of various three-dimensional figures

- The slope of a line and the slope-intercept form and point-slope form of the equation of a line

- The standard form of a quadratic equation as well as the quadratic formula

- The Pythagorean Theorem

- The formula for simple interest

Throughout this book, whenever one of these formulas is used, the text tells whether it is on the formula sheet. You are expected to know certain formulas that are not on the sheet. These are also identified in this book when they are used. The other foundational formulas that you are expected to know (and therefore are not included on the formula sheet) are the following:

- The total cost formulas that take into account markups and markdowns (see Chapter 3)

- The distance formula, $d = rt$ (see Chapter 5)

- Certain geometric formulas for perimeters, circumference, and areas of two-dimensional figures (see the table at the end of Chapter 7)

- The measures of central tendency, such as the mean, median, and mode (see Chapter 8)

Symbol Selector Sheet

For fill-in-the-blank questions, you may have to use symbols that are not readily available on the keyboard. To facilitate using symbols, the GED® test provides a symbol selector sheet as a drop-down item for those questions that require a symbol. When you choose a symbol, be sure to click on "Choose" and "Insert" from the bottom of the symbol selector sheet so that the symbol is included in your answer.

The symbol sheet contains symbols, but it does not explain what they mean. The table below includes the symbols as well as their meanings.

π	f	≥	≤	≠	2	3	||	×	÷	±	∞	√	+	−	(	)	>	<	=
pi, used with circles and spheres	function	greater than or equal to	less than or equal to	not equal to	exponent 2 (squared)	exponent 3 (cubed)	absolute value	multiply	divide	plus or minus	infinity	square root	plus	minus	open parenthesis	close parenthesis	greater than	less than	equals

Erasable Note Board

You will be given an erasable note board on which to do your calculations or make notes (see *gedtestingservice.com* and search for "noteboards"). You are not allowed to use the note board before the test begins. If you need an additional note board during the test, you can request a "fresh" one but you must turn in the one that is used. You may have only one note board at a time, and you must turn in the note board at the end of the test. The note board takes the place of scratch paper, so use it for that purpose. It is erasable.

The best advice for using the note board is to practice writing small so you don't have to get another note board. Erase when you are sure you won't need that information anymore. Before you even start with question 1 (but certainly after the test has begun), write (small) at the top of the note board anything you may need to jog your memory during the test. As examples, write PEMDAS to remind you of the order of operations, or the formulas for the circumference or area of a circle—whatever you think you might need and perhaps forget in the stress of taking the test. Leave these reminders on the note board for the duration of the test. That's why it's a good idea to write small—you'll still have space on the board to do calculations or sketch a figure.

Time Clock

A test-timer clock is located on the upper right-hand corner of the test display. You can minimize the clock or maintain it in view. However, during the final few minutes of the test, in order to make certain that you are aware that time is running out, the clock will appear and cannot be minimized.

It is a good idea to keep the clock in view to help you to pace yourself. The test allows a little less than two minutes per question, so if any multiple-choice or drop-down question seems to be taking more than a minute of your time, answer it to the best of your ability (eliminating obvious wrong choices), flag it for review (see next section), and go on to the next question. It might be a good idea to mark on your note board the test number and the answers that you should look at again, so when you go back you aren't rereading answers that you have already eliminated. Mark for review any other type of question that baffles you. Then, when you finish all of the questions, you should have enough time to go back to these questions.

Flag for Review

If you think you want to go back to any question, answer it to the best of your ability and then press the "Flag for Review" button to return to it later. If you can eliminate one or two choices, you can make an informed guess, but rather than taking too much time right then, flag that question, mark the number of the question and the remaining choices on your note board, go on to the next question, and come back later. Often, your brain is working in the background while you continue the test, still mulling over the ones you flagged, and when you return to them, the answers may come to you right away. You definitely do not want any unread questions at the end of the test just because you ran out of time by spending too much time on a question that momentarily baffled you. By the way, the question number (written in the form "Question 1 of 50," for example) is in the upper right corner above the Flag for Review option.

Flagging a question for review is an excellent idea. If it is taking more than a minute to figure out an answer, especially on a multiple-choice or drop-down question, mark your best guess (eliminating all of the obviously wrong options) and click on "Flag for Review" on the upper right side of the screen. When you have finished all of the questions, an Item Review Screen will appear, which will show which questions were answered and unanswered. The flag next to the question number will be filled in if you have flagged a question for review.

A menu at the right bottom of the screen will give you three choices: Review All, Review Unanswered, Review Flagged. If you choose Review Flagged, which you should at this point, it will take you through all of the flagged questions so you don't have to waste time looking for them. Of course, if you click any specific question on the Review Screen, that question will pop up on the screen as well. But the Review Flagged button will save some time because it presents the flagged questions only, one right after the other when you press the "Next" button on the bottom right of the screen. To answer the flagged questions, refer to your notes on the note board about which choices you should consider (these should have excluded the already eliminated ones). Those notes will save time, as stated above in the Time Clock section.

When you have finished answering all of the flagged questions, the Item Review Screen reappears, and now you can choose Review Unanswered. If you absolutely don't know an answer, just guess as best you can, click "Next," and go on to the next unanswered question. Do not leave any question unanswered. There is always a chance that you guessed the correct answer, which will improve your score. If you answer incorrectly or not at all, you get a zero for that question, so it is better to guess than to give up.

At the end of the test, click on "End Test" in the lower left corner of the Item Review Screen, if it indicates that all questions have been answered. If there are any that are still unanswered, click on them individually to fill in an answer before you click "End Test."

Practice Tests

This book, although paper-based, presents practice test questions in a format that is easily translatable to the computer-based format.

The GED® Testing Service provides a free sample test (not full-length) with limited functionality (go to *www.gedtestingservice.com* to find it). In addition, starting in November 2013, the official GED® practice test, called GED Ready™, will be available for a nominal fee (go to *www.gedmarketplace.com*, and be sure you get the 2014 version).

You are expected to be familiar with the testing environment before you take the actual test. Therefore, the GED® Testing Service provides a no-cost computer skills tutorial to be taken prior to the actual testing appointment. Go to *www.gedtestingservice.com* and search for "computer tutorials".

Calculator Instructions

If possible, get a scientific calculator, preferably the TI-30XS, which is the same as the one that is presented virtually on the actual test so you feel comfortable using it on the test. If cost is a consideration, search the Internet for used calculators. Go to *www.atomiclearning.com/ti30xs* for tutorials on how to use the TI-30XS. These show quite well "which buttons to press."

You can also download a free graphing calculator, such as the Mathlab Calculator app, to your cell phone, if it has that capability. Like the TI-30XS, this type of calculator has more functions than you will need for the GED® test, but it has the all-important ability to display fractions as fractions as well as decimals and to display square roots and absolute values exactly as they appear on the test (that is, it displays the square root symbol $\sqrt{25}$ in addition to the calculator-speak "sqrt25"). Practicing on a virtual calculator, even on your cell phone, will help to give you the dexterity and confidence you need to handle the calculator on the GED® test.

Overview

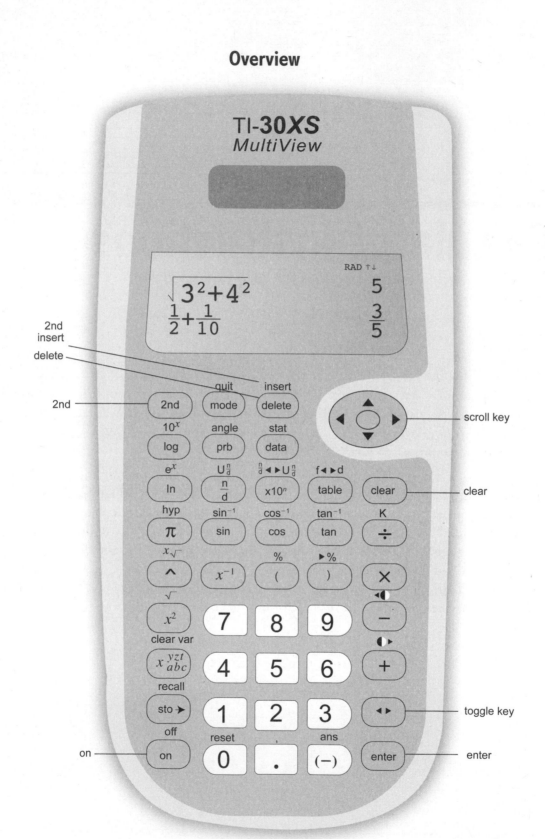

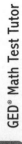

CALCULATOR

OVERVIEW OF TI-30XS CALCULATOR

Opposite each instruction sheet (as shown here) is a picture of the TI-30XS with the relevant keys circled to help you follow along. With practice, you should be able to use the calculator easily during the test. The keys circled on this first picture are the ones that are relevant to all functions. You won't need the (on), (mode), or (clear) keys during the test, but you will need them on a handheld calculator, so they are included here. Make sure the TI-30XS calculator is set to MathPrint mode before you start by selecting it via the (mode) key. You will need to set it only once, when you first start to use the calculator, and it will be stored in memory. The (clear) key clears the display so you can go on to the next problem.

You will need the rest of the keys displayed here for the GED® test virtual calculator as well as a handheld calculator. The (enter) key is the same as the equals key (=) on most other calculators. Use it to get the answer. The toggle key (<>) converts an answer from a fraction or square root to a decimal. To convert a fraction or square root that is not the last answer to a calculation, key it in and then press the (enter) key before the toggle key (<>).

To move the cursor on the calculator screen in any of the four directions, use the scroll button. The (delete) key deletes the character at the cursor, and (2nd) (delete) inserts a character at the cursor without deleting any of the other characters on the screen.

The (2nd) key acts like the shift (or ↑) key in text messaging or on the computer in that it converts the next keystroke to the value above it on the TI-30XS keyboard. For example, (2nd) x^2 gives the square root, and as we saw above, (2nd) (delete) becomes (insert).

Basic Arithmetic

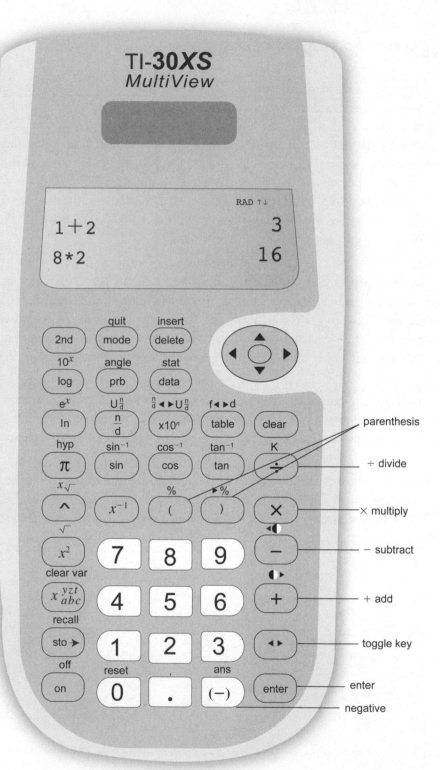

CALCULATOR

BASIC ARITHMETIC

The important keys for addition, subtraction, multiplication, and division are ⊕, ⊖, ⊗, ⊘, the parentheses keys ⦅ and ⦆ above the number pad, as well as (enter).

For **addition**, enter the first number, then ⊕, then the next number, etc., then (enter).

For **subtraction**, enter the first number, then ⊖, then the next number, etc., then (enter).

For **multiplication**, enter the first number, then ⊗, then the next number, etc., then (enter). The multiplication sign on the screen will change to ∗.

For **division**, enter the dividend (the number that is to be divided, or the numerator, the top number on a fraction), then ⊘, then the divisor (the number being divided into the dividend, or the denominator, the bottom number on a fraction), then (enter).

For any of the operations or any combination of operations, if any of the entries involve more than one term or factor, use parentheses or the answer might not be correct. For example, if you want to multiply $6 \times (2 - 5)$, enter it just that way, with the parentheses, and press (enter). The answer is ⁻18. The superscripted minus sign means it belongs to the number following it. If this calculation is entered without the parentheses, as 6 ⊗ 2 ⊖ 5, the answer is 7, which is incorrect.

If any numbers are negative numbers, use the ⊖ key on the keyboard for these numbers. If you enter 2 ⊗ ⊖ 6 using the ⊖ key, the screen will read 2∗⁻6 with ⁻12 as the answer. If you enter just 2 ⊗ ⊖ 6, the computer will return "SYNTAX error" because it reads the minus sign as the operation of subtraction.

For any answers, the toggle key ⟨⟩ will convert decimals to fractions and back again.

Fractions

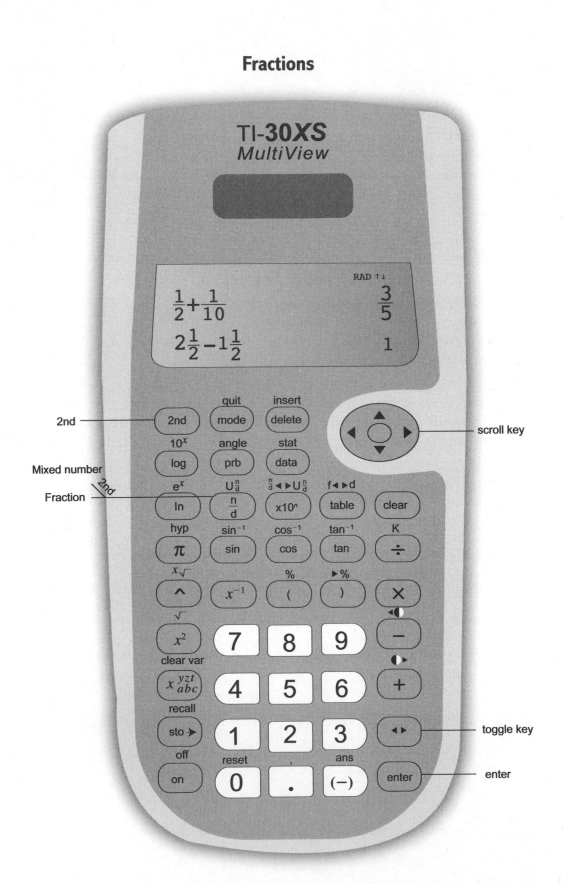

CALCULATOR

FRACTIONS

The important keys for calculating fractions are $\tfrac{n}{d}$, 2nd $\tfrac{n}{d}$, scrolling ⬦, the toggle key <>, and enter.

To enter a **fraction**, enter the numerator, press $\tfrac{n}{d}$, then enter the denominator. The screen will look like $\tfrac{\square}{\square}$. Scroll to the right ▷ to leave the fraction mode. For example, for $\tfrac{3}{4}$, press 3, $\tfrac{n}{d}$, then 4, scroll right ▷. Use the toggle key <>, then enter to display the fraction as a decimal.

Note that if the fraction has more than a single entry in either position, enclose the expression in parentheses ( ). For example, to evaluate $\tfrac{3 \div 5}{4}$, press (3 ÷ 5), $\tfrac{n}{d}$, then 4, scroll right ▷.

To enter a mixed number, use 2nd, then $\tfrac{n}{d}$. The screen will look like $\square\,\tfrac{\square}{\square}$. Enter the whole number part, scroll right ▷, then enter the fraction part as above. For example, for $2\tfrac{3}{4}$, press 2nd, then $\tfrac{n}{d}$, then 2, scroll right ▷, and continue for $\tfrac{3}{4}$ as above, but in this case scroll after the numerator (3) to enter the denominator.

To convert any fraction to a decimal, just use the toggle key <> and enter.

To convert an improper fraction to a mixed fraction, use 2nd, then $x10^n$. For example, to get the mixed fraction for $\tfrac{21}{5}$, press 21, then $\tfrac{n}{d}$, then 5, scroll right ▷, then 2nd $x10^n$, then enter. This function is not on the GED® Calculator Reference pull-down, but it could come in handy.

For all operations, when you want the answer, press enter, which acts as an equals sign.

Percentages

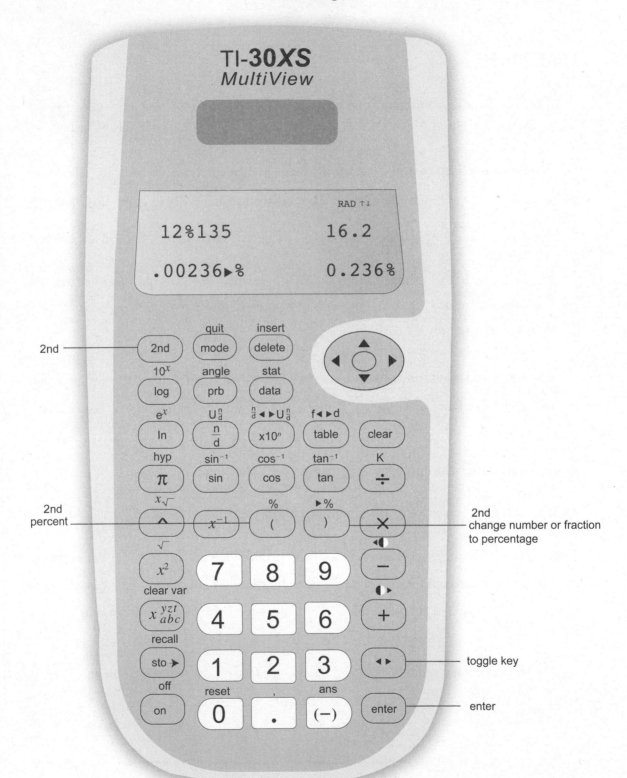

CALCULATOR

PERCENTAGES

The important keys for percentages are (2nd) (⊂) (which gives "% of"), the toggle key <>, and (enter), scrolling ◆, plus any of the basic operation keys.

For example, to find 12% of 135, enter 12, (2nd) (⊂), 135 and press (enter). The screen will look like 12%135 and will display the answer 16.2. Use the toggle key <> to change the decimal to a fraction.

To change a number to a percentage, use the (2nd) (⊃) combination. For example, to find the decimal .00236 expressed as a percentage, enter .00236 (2nd) (⊂) and press (enter). The result will show as 0.236%.

To display a fraction as a percentage, enter the fraction with the $\frac{n}{d}$ key, scroll right ▶, then (2nd) (⊃) and press (enter). For example, for $\frac{3}{4}$, press 3, then $\frac{n}{d}$, then 4, scroll right ▶, (2nd) (⊃), and press (enter). The result looks like $\frac{3}{4}$ ▸%, and the answer is 75%. This function is not on the GED® Calculator Reference pull-down, but it could come in handy.

Powers

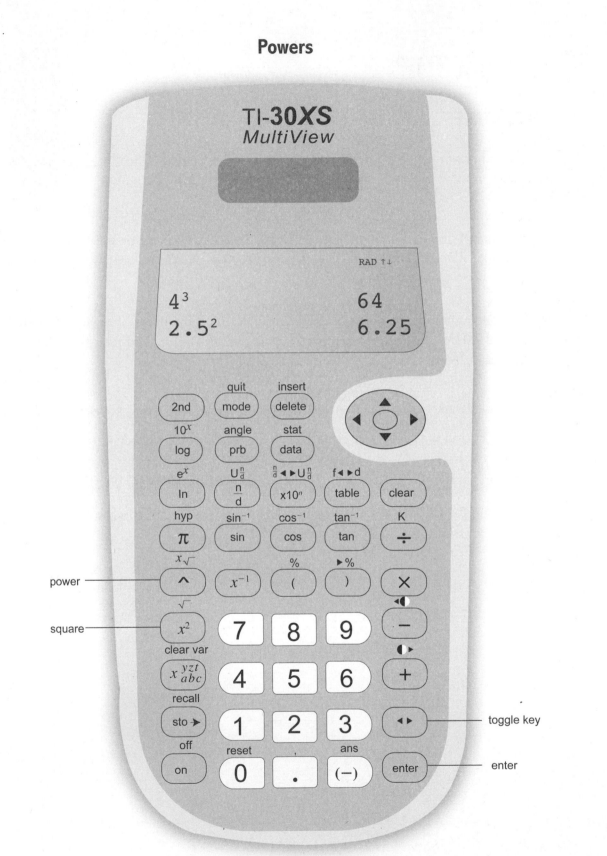

CALCULATOR

POWERS

The important keys for calculating **powers** are the x^2 and $\wedge$ keys, the toggle key $<>$, and (enter).

For example, to find 3^2, enter 3, then the x^2 key, and (enter). The screen will show 3^2 and give 9 as the answer. To find 4^3, enter 4, then the $\wedge$ key, then the exponent 3, and (enter). The screen will show 4^3 and give 64 as the answer. It seems like all that is needed is the $\wedge$ key because it does work for squares, but squares are so common, they get a "one-hit" key of their own. If you enter a decimal number, such as 2.5, its square is not a whole number (it is 6.25), but as for all decimal numbers, the toggle key $<>$ will convert it to a fraction.

Roots

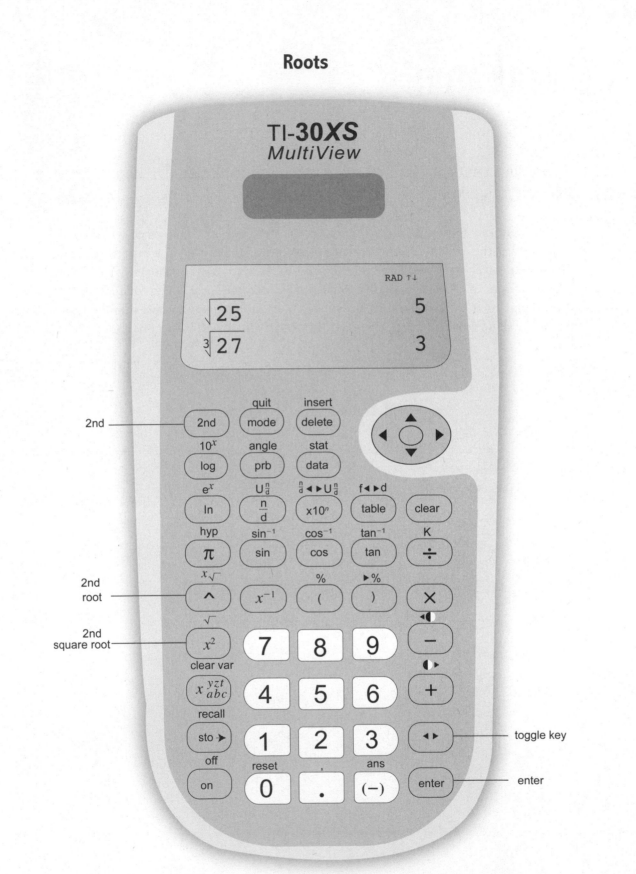

CALCULATOR

ROOTS

The important keys for calculating **roots** are the (2nd) (x^2) keys (for square roots) and the (2nd) (^) keys for other roots, the toggle key (<>), and (enter).

For square roots, enter the function first and then the number. For example, to find $\sqrt{25}$, enter the (2nd) (x^2) keys, 25, and (enter). The screen will show $\sqrt{25}$ and 5 as an answer.

If the root is not a whole number, such as $\sqrt{8}$, the answer will come up in radical form, but the toggle key (<>) will convert it to a decimal. So for $\sqrt{8}$, the screen looks like $\sqrt{8}$ and the answer is $2\sqrt{2}$, but the toggle key converts this to 2.828427.

For cube roots and higher, enter the root number before the function keys. For example, to find $\sqrt[3]{27}$, enter 3 (the root), then the (2nd) (^) keys, then 27 and (enter). It seems like all that is needed is the (2nd) (^) keys, and it does work for square roots, but square roots are so common, they get a key of their own.

Scientific Notation

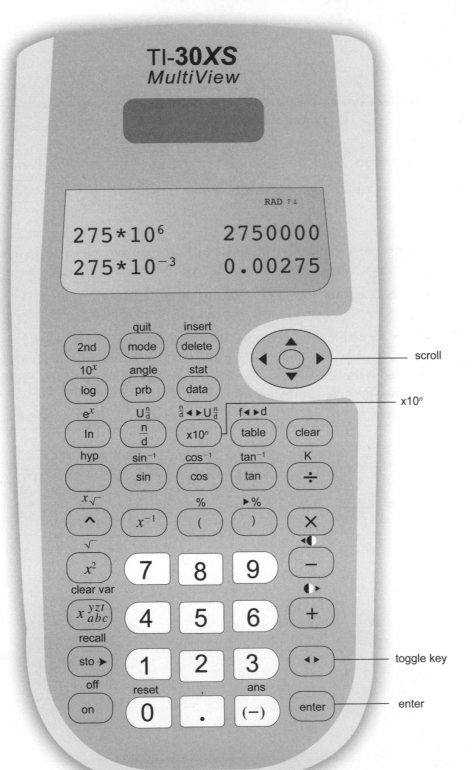

CALCULATOR

SCIENTIFIC NOTATION

The important keys for scientific notation are the $x10^n$ key, scrolling $\odot$, the toggle key $\boxed{<>}$, and $\boxed{\text{enter}}$.

For example, to change 2.75×10^6 to a whole number, enter 2.75, press the $x10^n$ key, 6 (the value for n), and $\boxed{\text{enter}}$. The result, 2750000 shows as the answer. Likewise, for 2.75×10^{-3}, enter 2.75, press the $x10^n$ key, $\boxed{(-)}$ 3 (the value for n), and $\boxed{\text{enter}}$. The result, 0.00275 shows as the answer. So it works for both positive and negative exponents, but you must use the $\boxed{(-)}$ key on the keyboard for negative exponents.

"I Was Born Knowing That"

Have you ever thought when asked "How do you *know* that?" that you were born knowing it. You cannot remember *how* you learned it, but as far back as you can remember, you always knew it. An example is the simple "one plus one equals two." You actually have a lot of math knowledge that you haven't even thought of. You know it, but you just don't *know* that you know it.

This book builds on your current knowledge to develop a mindset that promises to help you on many levels, such as everyday situations (checking to make sure you got the correct change when you buy something) and workplace problems (figuring what the deductions from your paycheck should be), as well as passing the Math Reasoning portion of the 2014 GED® test. All it involves is some insight and calling on the knowledge you already have.

Just as important as knowing math basics is knowing what to expect on the GED® test, so be sure you understand the information in Chapter 1. It is important to know how the test will be structured as well as to develop familiarity with the virtual calculator and other aids provided on the test (formula sheet, calculator reference, and symbol sheet).

This chapter starts with some basic math that you probably already know—again, you just *know* it—and builds on this information. For this chapter, try the examples near the beginning of each section first—if you think you already know the topic and can do all of the examples correctly, skip to the next section. Throughout the book, concentrate on learning the topics you aren't sure about. You can always go back if you forget something (use the index to find the page numbers).

CALCULATOR

BASIC ARITHMETIC

The important keys for addition, subtraction, multiplication, and division are $+$, $-$, $\times$, $\div$, the parentheses keys $($ and $)$ above the number pad, as well as enter.

For **addition**, enter the first number, then $+$, then the next number, etc., then enter.

For **subtraction**, enter the first number, then $-$, then the next number, etc., then enter.

For **multiplication**, enter the first number, then $\times$, then the next number, etc., then enter. The multiplication sign on the screen will change to *.

For **division**, enter the dividend (the number that is to be divided, or the numerator, the top number on a fraction), then $\div$, then the divisor (the number being divided into the dividend, or the denominator, the bottom number on a fraction), then enter.

For any of the operations or any combination of operations, if any of the entries involve more than one term or factor, use parentheses or the answer might not be correct. For example, if you want to multiply $6 \times (2 - 5)$, enter it just that way, with the parentheses, and press enter. The answer is ‾18. The superscripted minus sign means it belongs to the number following it. If this calculation is entered without the parentheses, as 6 $\times$ 2 $-$ 5, the answer is 7, which is incorrect.

If any numbers are negative numbers, use the $(-)$ key on the keyboard for these numbers. If you enter 2 $\times$ $(-)$ 6 using the $(-)$ key, the screen will read 2*‾6 with ‾12 as the answer. If you enter just 2 $\times$ $-$ 6, the computer will return "SYNTAX error" because it reads the minus sign as the operation of subtraction.

For any answers, the toggle key $<>$ will convert decimals to fractions and back again.

The GED® virtual TI-30XS calculator will be a time-saver on the test, but it is essential that you also study how to use it. If you don't know what the many keys mean, it won't be as useful to you and will take up your time instead of saving time. Be sure you are familiar with the calculator, even though you probably can do most of the problems in this chapter without a calculator. Some problems on the test (about 10%) will not allow you to use the calculator, so be prepared for that. In fact, it's a good idea to try to do the arithmetic in this chapter both ways: with and without a calculator.

Although the chapters in this book are divided into the topics on the GED® test, each chapter includes all of the information learned in the previous chapters. In addition, "calculator boxes" in the chapters, such as the one on the facing page, present how to use the virtual calculator for a particular subject area or type of problem. At the end of the instructional chapters, you will have all of the information you need to go on to do the practice tests confidently.

The Number Line

The above line is just a portion of the number line, which really goes on forever (to infinity, or ∞) in both the positive and negative directions. The **whole** numbers are marked on the line, but there are infinitely many numbers between each pair of whole numbers—for example, fractions and decimals, which we discuss in the next chapter. Numbers have various classifications, such as odd or even, and positive or negative.

The numbers get larger as we go to the right on the number line. That is true whether we consider positive numbers or negative numbers. Using the number line above and the symbols < for "less than" and > for "greater than," we see that it is true, although not always obvious, that

$$10 > 3 \qquad 3 > -1 \qquad -1 > -4 \qquad -10 < -5 \qquad -3 < 0 \qquad 5 < 9.$$

To remember which of the symbols, $<$ and $>$, is which, look at the left-hand sides. The "less than" sign ($<$) has the smaller (lesser) side on the left, and the "greater than" sign ($>$) has the bigger (greater) side on the left.

Example 2.1.

Rearrange the following numbers by size, starting with the smallest:

$$8 \quad 9 \quad -10 \quad 2 \quad 16 \quad -3 \quad -5 \quad 0 \quad -7 \quad 10$$

Answer 2.1.

$$-10 < -7 < -5 < -3 < 0 < 2 < 8 < 9 < 10 < 16$$

HINT

One way to remember negative numbers is to think of temperature: -10 degrees is colder (less warm) than -5 degrees, so $-10 < -5$.

Addition and Subtraction

Basics

Our number system is based on 10 (called the decimal system). It is believed that this is because the ancient mathematicians who "invented" the method of counting used their ten fingers (okay, eight fingers and two thumbs) to keep track when counting. The fact that we call the numbers 1 to 9 "digits" lends credibility to this theory because *digit* is another word for *finger*.

Numbers greater than 9 have two digits; the digit on the far right represents ones (also called units), and the next place to the left represents tens, then to the left of that is hundreds, and so on, as shown in the diagram. Note that commas are added to make it easier to read a large number.

3,	4	5	6,	7	8	9
millions	hundred thousands	ten thousands	thousands	hundreds	tens	ones

So the number 83,724 has 8 ten-thousands, 3 thousands, 7 hundreds, 2 tens, and 4 ones, and it is read as "eighty-three thousand, seven hundred twenty-four." Note that we say "eighty-three thousand" instead of "eight ten thousand, three thousand." We usually say the hundreds, tens, and ones together in each group of three numbers that are separated by the commas.

$$
\begin{array}{rccccc}
  &   &   &   &   & 4 \\
+ &   &   &   & 2 & 0 \\
+ &   &   & 7 & 0 & 0 \\
+ &   & 3 & 0 & 0 & 0 \\
+ & 8 & 0 & 0 & 0 & 0 \\
\hline
= & 8 & 3 & 7 & 2 & 4
\end{array}
$$

Note above that we filled in zeros to keep the columns straight. The zeros simply mean there are no units digits for the 2 tens, and no tens or units digits for the 7 hundreds. If you can't add larger numbers in your head, write them in column form. For any addition or subtraction problems, it is important to keep the digits in their correct columns. Of course, on the GED® test, you may be able to use the calculator.

When you hear the word *arithmetic*, what usually comes to mind right away is $1 + 1 = 2$, $2 + 2 = 4$, and so on. Or perhaps you also thought $3 - 1 = 2$. This is information you were "born knowing." For example, try the following problems—the correct answers follow.

Example 2.2.

Perform the following operations:

    a. 6 + 7 =

    b. 23 + 15 =

    c. 34 + 43 =

    d. 8 − 3 =

    e. 36 − 12 =

    f. 67 − 51 =

Answer 2.2.

    a. 13

    b. 38

    c. 77

    d. 5

    e. 24

    f. 16

If you got these correct, go to the next section. Otherwise, look again at the number line to get an idea of how we got these answers.

For example, for 6 + 7, start at 6 and move seven places to the right (when adding, move to the right). The answer is 13. If you have 6 pencils and you get 7 more, you now have 13 pencils.

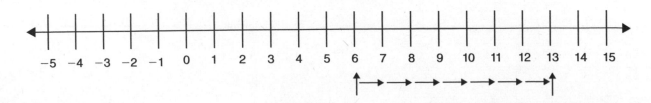

For $8 - 3$, which is the same as $8 + (-3)$, start at 8, but this time move three spaces to the left (when subtracting, or adding a negative, move to the left). Or think of "more" for addition and "less" for subtraction. If you start with 8 pens and you have 3 less, you now have 5 pens.

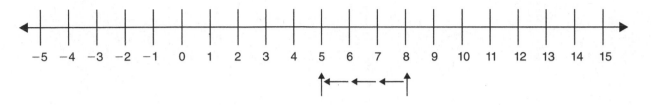

For problems involving more than one digit, write them in column form, and add each column separately:

$$\begin{array}{r} 23 \\ + 15 \\ \hline 38 \end{array}$$

Regrouping

Now we'll step it up a bit. Regrouping involves carrying and borrowing digits. Do the problems in the following example—if you get all of it right, skip to the next section.

Example 2.3.

Perform the following operations:

a. $13 + 29 =$

b. $57 + 55 =$

c. $75 + 36 =$

d. $34 - 16 =$

e. $73 - 59 =$

f. $82 - 79 =$

a. 42

b. 112

c. 111

d. 18

e. 14

f. 3

These problems involve carrying and borrowing. "Carrying" happens when the two numbers add to more than 9. Example 2.3(a) written in column form looks like this:

$$
\begin{array}{r}
13 \\
+\ 29 \\
\hline
42
\end{array}
$$

Start at the units column and work to the left. The units column has the numbers 3 and 9, for a total of 12, which has a 1 in the tens column and a 2 in the units column, so let's put the 1 where it belongs, in the tens column.

$$
\begin{array}{r}
1 \\
13 \\
+\ 29 \\
\hline
42
\end{array}
$$

This is the "1" from the fact that $3 + 9 = 12$.

Similarly, in subtraction, if you have to subtract a larger number from a smaller number in any column, you can "borrow" 1 from the column to the left, keeping in mind that the 1 really is 1 of whatever its original value was. Example 2.3(d), if written in column form, looks like this:

$$
\begin{array}{r}
34 \\
-\ 16 \\
\hline
18
\end{array}
$$

The units column has the numbers 4 and 6, but you have to subtract 6 from 4. If you borrow 1 from the tens column, that makes the 4 into a 14. Of course, it also changes the 3 in the tens column into a 2.

$$\begin{array}{cc} 2 & 14 \\ -\ 1 & 6 \\ \hline 1 & 8 \end{array}$$

The 3 in the tens column gave up 10 to the units column, so now we have 2 tens and 14 units, which has the same value as the original 3 tens and 4 units.

You can check your answer in a subtraction problem (appropriately called the "difference") by adding the answer to the number being subtracted. If you get the other number in the problem, it checks out; otherwise, find out where you made a mistake. For the example above, 16 + 18 indeed equals 34.

Positive Numbers, Negative Numbers, and Absolute Value

The word *absolute* in *absolute value* may make you think of a positive number. Actually absolute values are unsigned values. Try the following examples—if you get them all correct, skip this section and go on to the next.

Example 2.4.

Evaluate the following absolute values:

a. $|-3| =$

b. $|475| =$

c. $|2 - 2| =$

d. $|3 - 1| =$

e. $|1 - 3| =$

f. $|3 + 4 - 2 - 8| =$

    a. 3

    b. 475

    c. 0

    d. 2

    e. 2

    f. 3

Positive numbers don't have to have a sign in front of them. The number 8 means $+8$. Negative numbers *always* have to have a minus sign in front of them; otherwise, how would we know they aren't positive? Treat negative numbers the same as positive numbers in addition and subtraction with the following rule: keep track of the signs, specifically that minus a negative (two negative signs in a row) is a positive.

It is often helpful to put parentheses around a negative number until it is combined with the operation in front of it. For example, $8 - (-5)$ is the same as $8 + 5 = 13$.

( ) → "DO ME FIRST"

Parentheses mean "do me first." Putting negative numbers in parentheses reminds you to look at the numbers in the parentheses first and substitute their equivalent values. For example,

$$+ (-3) = -3 \quad \text{and} \quad - (-3) = +3.$$

Chapter 4 talks about the order of operations, for which this parentheses rule is very important.

As we have seen, numbers are really a measure on the number line of how many units each number is from 0. So 8 is 8 units from 0, and -5 is 5 units from 0 (on the other side of 0, but still 5 units). The difference, which measures the total distance from 8 to -5, is 13 units because there are 13 units between 8 and -5. The difference of the two numbers that are on the same side of zero is simply that—the difference in their distances. So $-8 - (-5) = -3$, which indicates 3 units on the negative side of 0. Note again how minus a negative becomes the addition of a positive: $-8 - (-5) = -8 + 5 = -3$, or 3 units on the negative side.

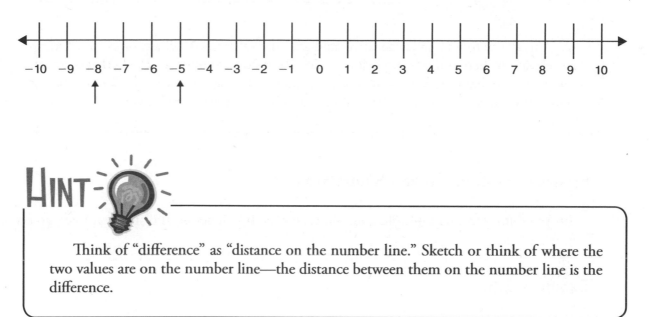

HINT

Think of "difference" as "distance on the number line." Sketch or think of where the two values are on the number line—the distance between them on the number line is the difference.

The distance of a number from 0 on the number line is actually its **absolute value**. It is the number of units, and it doesn't matter whether the number is positive or negative. In other words, the numbers $+8$ and -8 are each 8 units from 0 on the number line. Therefore, the distance from the point -8 to the point 8 on the number line is 16 units. We add their absolute values.

The symbol for absolute value is a pair of vertical lines | | that indicate (after anything between these lines is evaluated) the numerical part without regard to sign. Therefore, $|5| = 5$, $|-4| = 4$, and $|3 - 5| = 2$.

Of what use is absolute value in everyday life? If your bank balance is $-\$1,000$, you are $\$1,000$ in debt. This is because $|-1,000| = 1,000$. This number describes the amount of debt—you wouldn't say you are $-\$1,000$ in debt. The "in debt" takes care of telling whether the $\$1,000$ represents a positive or negative bank balance. A bank balance even less than $-\$1,000$, say, $-\$1,500$, says you are in even greater debt (talking about the absolute value). Again, the word *debt* tells that it is a negative balance. Now, suppose you got some money together to pay off the debt. Your deposit of $+\$1,500$ added to your present balance of $-\$1,500$ leaves a balance of $\$0$.

Properties of Addition and Subtraction

Try the following two multiple-question examples. If you get both of them correct, go on to the next section.

Example 2.5.

Which of the following addition problems is equivalent to 15 + 23?

    A. 23 + 15

    B. 8 + 7 + 23

    C. 10 + 13 + 15

    D. all of them

Answer 2.5.

(D) All of them. In addition, the order makes no difference, and even if we break the numbers up into their equivalents, we can switch the numbers around.

When doing a multiple-choice problem, don't choose a correct answer right away, such as (A) in Example 2.5, without at least glancing at the other answer choices. Even though (A) is equivalent to 15 + 23, so are the others, so (A) isn't the correct answer; (D) is.

Example 2.6.

Which of the following problems is equivalent to $23 - 15$?

    A. $15 - 23$

    B. $23 - 7 + 8$

    C. $23 - (7 + 8)$

    D. $7 + 8 - 23$

Answer 2.6.

(C) $23 - (7 + 8)$. The only correct answer is (C) since subtraction is not commutative (see next paragraphs). Parentheses mean "do me first," so $7 + 8 = 15$ and (C) becomes $23 - 15$. Answer choice (B) is different from (C) because without the parentheses, the operations are just done from left to right.

Addition has some very useful properties, which all come from the fact that you can add numbers in any order. This is called the **commutative property**. For example, $3 + 5 = 5 + 3$. This extends to adding any string of positive numbers: $3 + 5 + 6 = 5 + 3 + 6 = 6 + 3 + 5$, and so on, which also means that you can group the numbers in addition (the grouping has the name **associative property**) to make finding the sum easier. When we add $3 + 6 + 7 + 4$, it is faster to add the $3 + 7$ together and the $6 + 4$ together because we recognize $3 + 7 = 10$ and $6 + 4 = 10$, and the answer is 20. Compare that to adding in the order given: $3 + 6 = 9$, then $9 + 7 = 16$, then $16 + 4 = 20$. Whenever you can group numbers together that add to 10, it makes adding a string of numbers easier than adding them in whatever order they were given.

The operation of subtraction isn't commutative, so it doesn't have these properties. For example, $7 - 4 \neq 4 - 7$.

Multiplication

If you can do the following problems accurately (with pencil and paper but without a calculator), skip to the next section.

Example 2.7.

Perform the following operations:

a. $6 \times 9 =$

b. $7(8) =$

c. $12 \cdot 9 =$

d. $23 \times 4 =$

e. $(80)(6) =$

f. $23 \cdot 49 =$

Answer 2.7.

a. 54

b. 56

c. 108

d. 92

e. 480

f. 1,127

Multiplication can be indicated in four ways ($\times$, parentheses, *, and $\cdot$). They all mean the same thing: multiply the numbers. Note that no operation sign between a number and an opening parentheses, such as 3(6), also indicates multiplication.

When we write $2 \times 3 = 6$, we are actually saying 2 is added to itself three times. The whole numbers 2 and 3 are called **factors**, and the answer 6 is called the **product** of 2 and 3. Some numbers have several factors. For example, since $12 = 1 \times 12$, $12 = 2 \times 6$, and $12 = 3 \times 4$, all the factors of 12 are {1, 2, 3, 4, 6, 12}.

Some numbers do not have factors other than 1 and themselves. These are called **prime** numbers. Some examples of prime numbers are 2, 3, 5, 11, and 13. Can you think of the next prime number after 13? It's not 14 ($= 7 \times 2$), 15 ($= 3 \times 5$), or 16 ($= 2 \times 8$, or 4×4)—it's 17. No whole numbers multiplied together will give 17 except 1 and 17.

HINT

Other than the number 2, no even number can be a prime number because at least the number 2 is a factor. So, because it is even, 34 is not prime even though 17 is.

To do multiplication, your choices are (1) to memorize the multiplication table, (2) to figure out an easier way to remember or find the answers, or (3) to use a calculator. This last choice, using a calculator, is what more and more people are doing, and that is fine. But there are times when you don't have your calculator (which doubles as your cell phone—or the other way around), and on the GED® test, a small number of problems don't have the virtual calculator. What to do then?

You must become at least familiar with that multiplication table! That doesn't mean that you must memorize it, though. If you think about it, you were "born knowing" the 1's (of course, since anything multiplied by 1 is unchanged), the 2's (2, 4, 6, 8, . . .), the 5's (5, 10, 15, 20, . . .), and the 10's (10, 20, 30, 40, . . .). You also already know a lot of the others (3's, 4's, 6's, 7's, 8's, and 9's), but maybe not so readily.

If you have memorized the multiplication table, fine. But even then, some people get confused about a few of the values (for example, 6×9, 6×8, 7×8, 7×9; the answers are 54, 48, 56, 63). So let's look at a way we can figure it out based on what we know already (the 1's, 2's, 5's, and 10's). Because $3 = 2 + 1$, multiplying anything by 3 is the same as multiplying by 2 and then by 1 (easy enough), and than adding those numbers. For example, $7 \times 3 = (7 \times 2) + (7 \times 1) = 14 + 7 = 21$.

To multiply by 4, multiply by 2 and double it, so $7 \times 4 = (7 \times 2)$ doubled $= 14 \times 2 = 28$.

To multiply by 6, multiply by 3 and double it since $6 = 3 + 3$, so $7 \times 6 = 7 \times 3$ doubled, or $21 + 21 = 42$.

To multiply by 8, use $\times 10$ and then subtract $\times 2$ since $8 = 10 - 2$, so $7 \times 8 = (7 \times 10) - (7 \times 2) = 70 - 14 = 56$.

To multiply by 9, use × 10 and then subtract the number since 9 = 10 − 1, so 7 × 9 = (7 × 10) − 7 = 70 − 7 = 63.

To multiply by 7, there is no easy way, other than looking at the number it is being multiplied by and using the shortcut for that number.

This looks like more memorization, but it's all stuff you know and can figure out on your own. The important thing is that you can break things down to easier problems if you forget any part of the multiplication table. This is based on a property called the **distributive property**, discussed later in this chapter, which lets us do things like 16(3) = 16(2) + 16(1), or breaking one of the multipliers into smaller numbers and adding. Remember on the GED® test that you can use the erasable note board plus what you already know if the calculator is not available.

HINT

A quick way to remember the multiples of 9 is that the digits of each multiple add to 9 and the first digit is one less than the multiplier. Consider 6 × 9. One less than 6 is 5, the first digit, and since the first and second digits in the answer must add to 9, the second digit must be 4. So 6 × 9 = 54. Try it yourself.

Multiplying Numbers Greater Than 10

Now let's look at multiplying numbers with more than one digit. If you can do the following example without a calculator, skip this section and go on to the next.

Example 2.8.

Fill in the blanks. Do not use a calculator, but you may use paper and pencil.

a. 658 × 9 = [   ]

b. 16 × (−24) = [   ]

c. 242 × 11 = [   ]

a. 5,922

b. −384

c. 2,662

Let's multiply 824 × 4. Always start to multiply from the right side, or the ones column, so first we are looking at 4 × 4, which is 16.

$$
\begin{array}{r}
824 \\
\times\quad 4 \\
\hline
6
\end{array}
$$
The 6 comes from 4 × 4 = 16. We will "carry" the 1 (or 10) when we multiply 4 by the next column (the tens column).

$$
\begin{array}{r}
1 \\
824 \\
\times\quad 4 \\
\hline
96
\end{array}
$$
When we multiply 4 × 2 in the tens column, we get 8 in the tens column plus the 1 that we carried, so now we have 96 with no carrying.

$$
\begin{array}{r}
824 \\
\times\quad 4 \\
\hline
3,296
\end{array}
$$
When we multiply 4 × 8 in the hundreds column, we get 32, and we are done.

So multiplication, like addition, involves "carrying."

For larger numbers, such as multiplying 824 × 64, without a calculator, work it out as shown below. First multiply 824 by 4 and write that product, then multiply 824 by 6 (actually 60 because the 6 is in the tens column), write that product, and add the two results. This is really very much like the "breaking down" that we did above, just at a higher level.

$$
\begin{array}{r}
824 \\
\times\ 64 \\
\hline
3,296 \\
49,440 \\
\hline
52,736
\end{array}
$$
This is 824 × 4.
This is 824 × 60. Note the 0 placeholder in the ones column.
Add the two parts.

It is most important when doing "long" multiplication to keep the columns straight.

Multiplication of positive and/or negative numbers is the same as above, except that if the signs are the same, the answer is positive, and if the signs are different, the answer is negative. Thus,

$$4 \times 7 = 28 \qquad -4 \times 7 = -28 \qquad 4 \times -7 = -28 \qquad -4 \times -7 = 28.$$

Although it is important to know how to multiply numbers, and especially to be familiar with the product (answer) when multiplying one- or two-digit numbers, in real life as well as on the GED® test, a calculator will avoid careless errors.

Properties of Multiplication

This section discusses some of the most important facts about multiplication. They can simplify problems greatly. If you get Example 2.9 correct, skip this section and go to the next section.

Example 2.9.

Perform the following operations (with pencil and paper but without a calculator).

    a. $3 \times 6 \times 5 \times 4 =$

    b. $3(4 \times 5) \times 6 =$

    c. $3(4 + 5) \times 6 =$

    d. $4 \times 68 \times 25 =$

Answer 2.9.

    a. 360

    b. 360

    c. 162

    d. 6,800

Just as for addition, multiplication can be done in any order (again, the **commutative property**). So $3 \times 8 = 8 \times 3$. And $3 \times 5 \times 2$ is the same as $3 \times 2 \times 5$ or $5 \times 3 \times 2$ or any other combinations of these three numbers.

This property allows us to "group" numbers in multiplication (the grouping has the name **associative property**) to make finding the product easier. If we are asked to multiply $25 \times 7 \times 4$, and we recognize that $25 \times 4 = 100$, the problem is really the same as $7 \times 100 = 700$. Compare that to multiplying 25×7 and then multiplying 175×4, or worse yet, $25 \times (7 \times 4) = 25 \times 28$. They all give the same answer, but the first solution is much easier.

Another property of multiplication is that $6(3 + 4) = (6 \times 3) + (6 \times 4)$. This is called the **distributive property** because multiplication by 6 is "distributed" to each number being added in the parentheses. Of course, parentheses say "do me first," so this problem is more easily done as $6 \times 7 = 42$.

Sometimes you can recognize a common factor in an addition problem for which the reverse of the distribution property presented above can be useful. For example, you need to find the sum of 56 and 24. If you recognize that both of these numbers are multiples of 8 (said another way, 8 is a factor of both numbers) you could think: $56 + 24 = 8(7) + 8(3) = 8 \times (7 + 3) = 8 \times 10 = 80$.

The distributive property of multiplication plays a big role when we do algebra (see Chapters 5 and 6).

Division

If you can do the following problems accurately (with pencil and paper but without a calculator), skip to the next section.

Example 2.10.

a. $48 \div 6 =$

b. $56/8 =$

c. $\dfrac{864}{2} =$

d. $-72 \div 9 =$

e. $4\overline{)484} =$

f. $\dfrac{6{,}432}{16} =$

a. 8

b. 7

c. 432

d. −8

e. 121

f. 402

In the problems above, division is indicated in three ways (÷, as a fraction, and with $\overline{)}$). They all mean the same thing. We discuss fractions in more detail in Chapter 3.

Division is the "opposite" of multiplication, so if you know multiplication, you know division. The answers to Examples 2.10 (a), (b), and (d) show that. Remember that the ÷ sign is read as "divided by," so there should be no confusion—the first number (dividend) is divided by the second number (divisor). As a check, the quotient (answer) multiplied by the divisor should equal the dividend.

Although it is important to know how to divide numbers, in real life as well as on the GED® test, a calculator will avoid careless errors.

Long Division

If you don't have a calculator for larger numbers, such as 832 ÷ 16, set it up for long division:

$$16\overline{)832}$$

The first step is to divide the divisor (16) into the first digit of the dividend (8). It doesn't go into 8, so include the next digit: 16 does go into 83 (although there will be something left over). Using your best guess, you see that 16 divides into 83 five times. So the first number of the answer (quotient) is a 5. Be sure to align the 5 over the 3, not the 8—it is the quotient of 83, not 8. So you have

$$5$$
$$16\overline{)832}$$

Now multiply the quotient by the divisor ($5 \times 16 = 80$) and place this under the 83. Subtract and bring down the next digit (2). The calculation now looks like this:

$$5$$
$$16\overline{)832}$$
$$80$$
$$\overline{32}$$

Now, divide the 16 into 32, which goes 2 times, and the 2 goes in the quotient above the 2 in the dividend. When the 2 in the quotient is multiplied by the divisor, the answer is 32. And since $32 - 32 = 0$, there is no remainder. The final long division looks like this:

$$52$$
$$16\overline{)832}$$
$$80$$
$$\overline{32}$$
$$32$$
$$\overline{0}$$

We can check that the answer is correct by multiplying the quotient (52) by the divisor (16), and indeed we see that $52 \times 16 = 832$.

But what if you aren't sure how many times 16 goes into 83 in the above problem? You actually use your best guess, but let's say that guess was 6, not 5. Then the first step of the long division would look like this:

$$6$$
$$16\overline{)832}$$
$$96$$

and since 96 > 83, your guess of 6 is too much. Likewise, if your best guess is that 16 goes into 83 four times, the long division would start like this:

$$4$$
$$16\overline{)832}$$
$$64$$
$$\overline{19}$$

Because 19 is larger than 16, 16 can go into 83 at least one more time. The answer after subtraction must be smaller than the divisor.

Long division looks like some guesswork, and in truth it often is, but with a knowledge of multiplication, the guesses aren't wild guesses—they have some logic behind them. Again, most long divisions are done by calculator; however, it is important to know how to do long division as a backup.

> Note that we cannot divide by 0. Anything that indicates division by 0, such as $\dfrac{25}{4-4}$, is undefined.

Division of positive and/or negative numbers is the same as above, except that if the signs are the same, the answer is positive, and if the signs are different, the answer is negative. Thus,

$$28 \div 7 = 4 \qquad -28 \div 7 = -4 \qquad 28 \div -7 = -4 \qquad -28 \div -7 = 4$$

Not all quotients are whole numbers (with no remainder), like the problems above. The remainder indicates what is "left over," expressed either as a decimal or a fraction, which are topics of Chapter 3.

Properties of Division

Division is not commutative. In other words, $6 \div 3$ is not the same as $3 \div 6$, so some of the things we did to make multiplication easier aren't possible with division.

Exercises

(Do these with pencil and paper but without a calculator.)

1. Fill in the missing numbers:

    a. $3 + \boxed{} + 20 = 25$

    b. $3 \times 5 \times 20 = \boxed{}$

    c. $12 \div \boxed{} = 4$

    d. $17 - 7 + 5 = \boxed{}$

2. Which of the following groups of numbers are all prime numbers?

    A. 3, 7, 11, 21

    B. 5, 13, 17, 24

    C. 3, 17, 19, 23

    D. 2, 5, 9, 15

3. $(3 + 6) \times 2$ $\left\{\begin{array}{l} \text{is less than} \\ \text{is equal to} \\ \text{is greater than} \end{array}\right\}$ $3 + (6 \times 2).$

4. After surgery, Joe has to take a pain pill every 45 minutes. He takes the first one at 6:00 p.m. At what time should he take his next pill?

    A. 6:45 a.m.

    B. 6:45 p.m.

    C. 5:15 p.m.

    D. 7:30 p.m.

5. Which of the following is the same as $- |-7 - (-2)|$?

    A. $- | 7 + 2 |$

    B. $- | -7 + 2 |$

    C. $| -7 - 2 |$

    D. $| 7 + 2 |$

6. Jim rented a moving van at 9 a.m. on Monday and returned it at 5 p.m. on Tuesday. Since it was after 3 p.m. on Tuesday, he had to pay for two full days. His bill was $96.00. The daily rate was $ [   ] .

7. The expression $17 - (-5 + 9)$ is $\left\{\begin{array}{l} \text{less than} \\ \text{equal to} \\ \text{greater than} \end{array}\right\}$ $13.$

8. If four watermelons cost $12, the cost of one watermelon is [   ] .

9. Which is the best buy?

    I. $10 for 5 pounds of apples

    II. $6 for 3 pounds of apples

    III. $8 for 4 pounds of apples

    A. I

    B. II

    C. III

    D. They are all the same

10. Paint costs $29 for a one-gallon can and $112 for a five-gallon can. Roger needs four gallons for a job. The better deal is $\left\{\begin{array}{c}\text{four one-gallon cans} \\ \text{one five-gallon can} \\ \text{no difference}\end{array}\right\}$.

11. All of the factors of 24 are

    A. {2, 3, 4, 6, 8, 12}

    B {1, 2, 3, 4, 6, 8, 12}

    C. {1, 2, 3, 4, 5, 6, 12}

    D. {1, 2, 3, 4, 6, 8, 12, 24}

12. Luis worked 10 hours on Monday and was paid $9 per hour for the first 8 hours and double-time ($18 per hour) for the rest of the time. How much did Luis earn on Monday?

    A. $72

    B. $180

    C. $108

    D. $90

13. Carlton wants to buy a used car. He has to put $1,200 down. He can get a second job working 4 hours a night for $12 per hour. In how many nights will he have earned the down payment?

    A. 20

    B. 24

    C. 25

    D. 48

14. In expanded notation, the number 5,624 is $\boxed{}$ × 1,000 + 6 × $\boxed{}$ + $\boxed{}$ × $\boxed{}$ + 4.

15. Evaluate: $(-4)\,(-3)\,(1)\,(-1)\,(0)\,(10)\,(-2)$.

    A. 240

    B. −240

    C. 120

    D. 0

Solutions

Answer 1.   a. 2

   b. 300

   c. 3

   d. 15

Answer 2.   (C) 3, 17, 19, 23. In answer choice (A), $21 = 3 \times 7$. In answer choice (B), $24 = 3 \times 8$. In answer choice (D), $9 = 3 \times 3$ and $15 = 3 \times 5$.

Answer 3.   Is greater than. $3 + 6 \times 2 = 9 \times 2 = 18$. Parentheses say "do me first," so $3 + (6 \times 2) = 3 + 12 = 15$.

Answer 4.   (B) 6:45 p.m. This is 45 minutes after 6:00 p.m. Don't make the mistake of picking the first answer without seeing that it is a.m. instead of p.m.

Answer 5.   (B) $-|-7 + 2|$. Doing the parentheses first in the original expression yields $-|-7 - (-2)| = -|-7 + 2|$, which is answer choice (B). Choices (C) and (D) can be eliminated immediately because the original expression is minus an absolute value, so it must be a negative.

Answer 6.   48.00. The only relevant information in this problem is that Jim was charged for two days and the bill was $96.00. Sometimes a problem gives information that is not needed for the solution. Note that the answer is 48.00 and not $48.00 because the dollar sign is already provided.

Answer 7.   Equal to. Be sure to evaluate the parentheses first, giving $17 - 4 = 13$.

Answer 8. $3. Divide $12 by 4: $12 \div 4 = 3.

Answer 9.   (D) They are all the same. $10 \div 5 = 2. $6 \div 3 = 2. $8 \div 4 = 2. The cost per pound is $2 for all of them.

Answer 10.   One five-gallon can. The cost of four one-gallon cans at $29 each costs $4 \times $29 = 116. The five-gallon can costs less, at $112.

Answer 11. (D) {1, 2, 3, 4, 6, 8, 12, 24}. Don't forget to include the numbers 1 and 24 in the factors.

Answer 12. (C) $108. For the first 8 hours, Luis was paid 8 $\times$ $9 = $72. For the rest of the time (10 hours total minus the first 8 hours = 2 hours), he was paid $18 per hour, or 2 $\times$ $18 = $36. So in total he earned $72 + $36 = $108.

Answer 13. (C) 25. Each night Carlton will make 4 $\times$ $12 = $48. Therefore, the problem is to find $1200 $\div$ $48 to determine how many nights it will take.

Answer 14. 5; 100; 2; 10. In expanded notation, the number 5,624 is 5,000 + 600 + 20 + 4, or in expanded form, 5 $\times$ 1,000 + 6 $\times$ 100 + 2 $\times$ 10 + 4.

Answer 15. (D) 0. Zero times anything is 0, so before you multiply any long group of numbers, look to see whether one of them is a 0.

The Parts of the Whole

Decimals, fractions, and percentages are all closely related. For example, to say that 5 is half of 10 can be represented as $.50 \times 10$, $\frac{1}{2}$ of 10, or 50% of 10. This chapter discusses each of these types of calculations in detail.

Decimals

The four operations on decimals are essentially the same as for whole numbers with special attention given to the placement of the decimal point in the answer. Zeros can be filled in as placeholders to the left of a whole number (00123 is the same as 123), or to the right of a decimal (.45600 is the same as .456).

The word *decimal* comes from the Latin word that means "10." Our counting system (as we saw in Chapter 2) as well as our monetary system are based on the number 10. Just as our placeholders were ones, tens, hundreds, and so forth, for whole numbers, decimals indicate parts of units, with placeholders of tenths, hundredths, thousandths, and so on. The decimals appear after a decimal point (.) and get smaller as the numbers go to the right.

CALCULATOR

BASIC ARITHMETIC

The important keys for addition, subtraction, multiplication, and division are ⊕, ⊖, ⊗, ⊘, the parentheses keys ⊂ and ⊃ above the number pad, as well as (enter).

For **addition**, enter the first number, then ⊕, then the next number, etc., then (enter).

For **subtraction**, enter the first number, then ⊖, then the next number, etc., then (enter).

For **multiplication**, enter the first number, then ⊗, then the next number, etc., then (enter). The multiplication sign on the screen will change to *.

For **division**, enter the dividend (the number that is to be divided, or the numerator, the top number on a fraction), then ⊘, then the divisor (the number being divided into the dividend, or the denominator, the bottom number on a fraction), then (enter).

For any of the operations or any combination of operations, if any of the entries involve more than one term or factor, use parentheses or the answer might not be correct. For example, if you want to multiply $6 \times (2 - 5)$, enter it just that way, with the parentheses, and press (enter). The answer is $^-18$. The superscripted minus sign means it belongs to the number following it. If this calculation is entered without the parentheses, as 6 ⊗ 2 ⊖ 5, the answer is 7, which is wrong.

If any numbers are negative numbers, use the ⊖ key on the keyboard for these numbers. If you enter 2 ⊗ ⊖ 6 using the ⊖ key, the screen will read 2*⁻6 with $^-12$ as the answer. If you enter just 2 ⊗ ⊖ 6, the computer will return "SYNTAX error" because it reads the minus sign as the operation of subtraction.

For any answers, the toggle key ⊂⊃ will convert decimals to fractions and back again.

For a mixed number (whole number plus a decimal part), the columns have the following labels:

1	2	3	4	5	.	6	7	8	9
ten thousands	thousands	hundreds	tens	ones	decimal point	tenths	hundredths	thousandths	ten thousandths

The mixed number 365.421 is read as "three hundred sixty-five and four hundred twenty-one thousandths." The decimal point is read as "and," and the decimal portion is read as a whole number (421 here) with the designation of the smallest measure (thousandths). The expanded notation of the decimal portion of this number is 4 tenths, 2 hundredths, and 1 thousandth.

Adding and Subtracting Decimals

When adding or subtracting decimals, be sure all of the decimal points are aligned. Use zeros if necessary to keep the columns straight. Then add or subtract as with whole numbers, keeping the decimal point in the same place in the answer as it is for the numbers above.

For example, to add 82.36 + 3.945 + 600.2, or subtract 29.9 from 756.802, we set these up as:

$$
\begin{array}{r}
82.360 \\
3.945 \\
+\,600.200 \\
\hline
686.505
\end{array}
\qquad
\begin{array}{r}
756.802 \\
-\,29.900 \\
\hline
726.902
\end{array}
$$

Note that zeros are filled in on the numbers on the right side of the decimal point, but are unnecessary on the left side because we are used to adding whole numbers just by aligning the columns. We could have just aligned the columns on the right side, too, but we're not used to that, so we fill in the zeros to help keep the alignment.

Multiplying Decimals

Multiplication of decimal numbers is the same as for whole numbers, except we have to figure out where the decimal point goes in the product (the answer). Simply count all the numbers after the decimal point in the numbers being multiplied, and starting from the last number on the right, count back that many places and put the decimal point there. It's that easy. It is not necessary to align the decimal points of the numbers being multiplied when doing multiplication.

Therefore, 54.21×1.3 is set up as

54.21	The decimal point is at 2 places from the right.
$\times$ 1.3	The decimal point is at 1 place from the right.
16263	This is 5421×3. Note that we ignore the decimal points for now.
54210	This is 5421×10. Note the 0 placeholder in the last column.
70.473	Now we put the decimal point at 3 spaces from the right.

Since the decimal point was two spaces from the right in 54.21 and one space from the right in 1.3, it is placed (since $1 + 2 = 3$) three spaces from the right in the answer: 70.473.

Dividing Decimals

Division of decimal numbers is the same as for whole numbers, except again we have to figure out where the decimal point goes in the quotient (the answer). This is a little trickier than for multiplication. Set the problem up as long division the usual way. For example, for $96.39 \div 3.5$, we would have $3.5\overline{)96.39}$, and we proceed to divide as we did with whole numbers, ignoring the decimal points (so we proceed to just divide like we would for $35\overline{)9639}$). Now we place the decimal in the answer by the following method: Count the number of places after the decimal point in the divisor (here, there is one), and starting at the existing placement of the decimal point in the dividend, count over that many places (here, that location is between the 3 and the 9), and place the decimal point in the quotient (the answer) there. Here, that is between the 7 and 5, the numbers above the 3 and the 9. When done, the problem will look like the following:

$$
\begin{array}{r}
2\,7.54 \\
3.5{\uparrow}\overline{)96.3{\uparrow}90} \\
70 \\
\overline{263} \\
245 \\
\overline{189} \\
175 \\
\overline{140} \\
140 \\
\end{array}
$$

Note that we added the last zero in the dividend because adding zeros to the right after the decimal point does not change the value of a number, and in this case, it makes the answer come out evenly.

Rounding

Rounding an answer comes up frequently with decimals, although any number can be rounded. As an example, when something costs $24.95, we often say it costs $25. In this case, we rounded off to the nearest unit (dollar).

The method for rounding is to look at the next digit after the place value that we want to round. If it is less than 5, we just drop that digit and all the ones to the right (inserting zeros if necessary). If it is 5 or more, we add "1" to the digit to be rounded. This method works whether the number is a whole number or a decimal number. You just have to know which digit is being rounded.

So, for example, 1,346 rounded to the nearest hundred would be 1,300 because we look at the 4—the digit to the right of 3, the "hundreds" digit. Since 4 < 5, we leave the 3 alone and fill in zeros for the rest of the placeholders.

As another example, the mixed number 245.829

> to the nearest hundred<u>th</u> would give 245.83
>
> to the nearest ten<u>th</u> would give 245.8
>
> and to the nearest hundred would give 200.

 **HINT**

Be sure to round off by looking at the digit only one place to the right of the one being rounded. A common mistake in rounding off is to start at the rightmost digit instead of the digit just one place to the right of the one we are rounding.

For example, to round 1,346 to the nearest hundred, if we started at the 6 and rounded the 4 up to 5 (since $6 \geq 5$), we would have 1,350, and then the answer we would get for the nearest hundred would be 1,400, not the correct answer of 1,300.

Therefore, when dividing decimals, we carry out the division to the next digit after the one we want to round to. For example, if we were dividing $125.32 by 7, the answer would be

$17.90285714 (calculator answer); however, we would look only at $17.902 (ignoring the other numbers) and round to $17.90 because we would normally round money to the nearest penny (one-hundredth of a dollar). If the question had asked for the answer to the nearest dollar, we would carry the division only to $17.9 and then round up to $18.

Example 3.1.

Perform the following operations either by long division or by using a calculator. Round your answers to the nearest thousandth.

        a. $52.4 \times 8.12 =$

        b. $1.275 \times 3.49 =$

        c. $12 \div 7 =$

        d. $34.266 \div 4.2 =$

Answer 3.1.

        a. 425.488

        b. 4.450. Notice that the answer is 4.44975. To the nearest thousandth adds 1 to the "9" because the next digit over, the 7, is greater than 5. When 1 is added to 9, that changes the ".449" to ".450".

        c. 1.714

        d. 8.159

Example 3.2.

Ilana goes to the store and buys milk ($2.79), bread ($1.88), and two pounds of bananas (at 59 cents a pound). She pays the cashier with a $10 bill. The amount of change she will receive is

        A. $5.26

        B. $4.74

        C. $5.85

        D. $4.15

(D) $4.15. Answer choices (A) and (B) are, respectively, the total and the change if you didn't notice that there are *two* pounds of bananas at 59 cents a pound. Answer choice (C) is the total bill, and answer choice (D) is the change from a $10 bill, which is what the problem asked. Read all problems carefully!

Fractions

Fractions are more involved than working with decimals or whole numbers. Fortunately, there are shortcuts in several of the fraction operations that make things easier.

A fraction is made up of two parts, the **numerator** and the **denominator**:

$$\text{fraction} = \frac{\text{numerator}}{\text{denominator}}.$$

A **fraction** tells us how many parts of a whole thing. For example, the fraction $\frac{3}{4}$ means three parts of something that has four total parts. A **mixed number** has a whole number and a fraction, such as $2\frac{4}{5}$, which means 2 plus $\frac{4}{5}$, and is spoken as "two and four-fifths."

The calculator available on the GED® test has a function for fractions that makes the calculations easier (and more accurate). However, understanding fractions is essential for using the calculator to get the right answers.

The reason fractions look like division problems is because that is actually what they are. We already saw how to get a decimal answer by dividing two numbers, so we just have to remember which part of the fraction gets divided into which.

CALCULATOR

FRACTIONS

The important keys for calculating fractions are $\frac{n}{d}$, (2nd) $\frac{n}{d}$, scrolling ◆, the toggle key <>, and (enter).

To enter a **fraction**, enter the numerator, press $\frac{n}{d}$, then enter the denominator. The screen will look like $\frac{\square}{\square}$. Scroll to the right (▸) to leave the fraction mode. For example, for $\frac{3}{4}$, press 3, $\frac{n}{d}$, then 4, scroll right (▸). Use the toggle key <>, then (enter) to display the fraction as a decimal.

Note that if the fraction has more than a single entry in either position, enclose the expression in parentheses (( )). For example, to evaluate $\frac{3 \div 5}{4}$, press ((3 ÷ 5)), $\frac{n}{d}$, then 4, scroll right (▸).

To enter a mixed number, use (2nd), then $\frac{n}{d}$. The screen will look like $\square\frac{\square}{\square}$. Enter the whole number part, scroll right (▸), then enter the fraction part as above. For example, for $2\frac{3}{4}$, press (2nd), then $\frac{n}{d}$, then 2, scroll right (▸), and continue for $\frac{3}{4}$ as above, but in this case scroll after the numerator (3) to enter the denominator.

To convert any fraction to a decimal, just use the toggle key <> and (enter).

To convert an improper fraction to a mixed fraction, use (2nd), then (x10ⁿ). For example, to get the mixed fraction for $\frac{21}{5}$, press 21, then $\frac{n}{d}$, then 5, scroll right (▸), then (2nd) (x10ⁿ), then (enter). This function is not on the GED® Calculator Reference pull-down, but it could come in handy.

For all operations, when you want the answer, press (enter), which acts as an equals sign.

HINT

A quick reminder of which way to divide in a fraction is to think of the fraction for a quarter $\left(\dfrac{1}{4}\right)$ and a quarter in money (.25). Then you can remember that the 4 is divided into the 1 (denominator into numerator) because the other way around, 1 divided into 4, would give $4, lots more than a quarter!

Comparing Fractions

If we have two fractions with the same denominators, it isn't difficult to figure out which is the larger fraction. For example, we instinctively know that $\dfrac{1}{4} < \dfrac{3}{4}$. We can even visualize fractions: if we cut a pizza into 8 slices, we can visualize that 5 slices are more than 3 slices, or $\dfrac{5}{8} > \dfrac{3}{8}$.

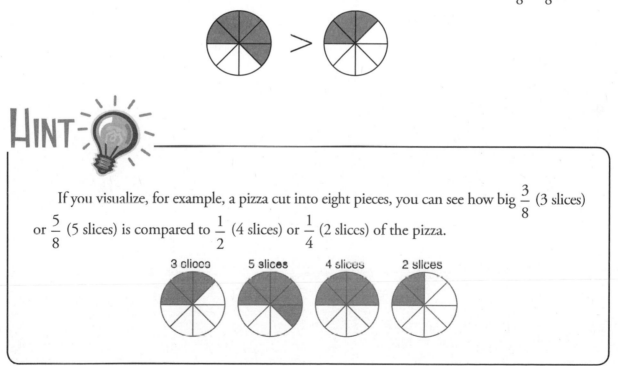

HINT

If you visualize, for example, a pizza cut into eight pieces, you can see how big $\dfrac{3}{8}$ (3 slices) or $\dfrac{5}{8}$ (5 slices) is compared to $\dfrac{1}{2}$ (4 slices) or $\dfrac{1}{4}$ (2 slices) of the pizza.

3 slices   5 slices   4 slices   2 slices

Also, often we can compare two mixed numbers by comparing just the whole number parts, such as $3\frac{1}{3} > 2\frac{2}{3}$. Even though the fraction parts $\frac{1}{3} < \frac{2}{3}$, the whole number 3 is clearly more than 2.

Fractions that have different denominators are not as easy to compare. For example, is $\frac{5}{7}$ greater or less than $\frac{7}{10}$? We need a quick way to check this out. Since decimals are quickly comparable, if we convert these fractions to their decimal equivalents, we will have our answer. Fortunately, the calculator gives the decimal equivalent of each fraction just by pressing (enter) and then the toggle (<>) key. Or we can just divide the dividend (top) by the divisor (bottom).

So we see that we get the decimal equivalent of $\frac{5}{7}$ by dividing the 5 by the 7 ($5 \div 7$), which will give us .7143. Since we know that $\frac{7}{10} = .7000$, we know that $\frac{5}{7} > \frac{7}{10}$.

Example 3.3.

Choose the appropriate sign for these fraction pairs:

a. $\dfrac{3}{7} \left\{ \begin{array}{c} > \\ = \\ < \end{array} \right\} \dfrac{5}{6}$

b. $\dfrac{3}{4} \left\{ \begin{array}{c} > \\ = \\ < \end{array} \right\} \dfrac{3}{8}$

c. $\dfrac{2}{9} \left\{ \begin{array}{c} > \\ = \\ < \end{array} \right\} \dfrac{5}{12}$

d. $\dfrac{4}{7} \left\{ \begin{array}{c} > \\ = \\ < \end{array} \right\} \dfrac{8}{14}$

e. $2\dfrac{11}{13} \left\{ \begin{array}{c} > \\ = \\ < \end{array} \right\} 2\dfrac{11}{26}$

a. $<$

b. $>$

c. $<$

d. $=$

e. $>$

Multiplication of Fractions

Even though addition and subtraction are easier operations for whole numbers and decimals than they are for fractions, *multiplication* of fractions is easier, so we will start with that. Division of fractions is just like multiplication, so if you understand multiplication of fractions, you also understand division.

Generally, when you multiply two or more fractions, you multiply the numerators to get the numerator of the answer, and then you multiply the denominators to get the denominator of the answer. For example, $\frac{4}{5} \times \frac{3}{4} \times \frac{6}{7} \times \frac{7}{8} = \frac{4 \times 3 \times 6 \times 7}{5 \times 4 \times 7 \times 8}$. However, three facts that we already know make this calculation easier:

1. Any number divided by itself equals 1, or, for example, $\frac{3}{3} = 1$.

2. Any number multiplied by 1 is unchanged, so $4 \times 1 = 4$.

3. The value of 3×4 is the same as 4×3 (the **commutative property** of multiplication).

If we multiply the above fractions out, $\frac{4}{5} \times \frac{3}{4} \times \frac{6}{7} \times \frac{7}{8} = \frac{4 \times 3 \times 6 \times 7}{5 \times 4 \times 7 \times 8}$, we get as a final answer $\frac{504}{1120}$. But let's switch some of the numbers around. We already know that it won't affect the value of the final fraction. So let's get the equivalent $\frac{4 \times 3 \times 6 \times 7}{4 \times 5 \times 8 \times 7}$ by switching the 5×4 and the 7×8 in the denominator to 4×5 and 8×7. Since $\frac{4}{4} = 1$ and $\frac{7}{7} = 1$, this is the same as $1 \times \frac{3 \times 6}{5 \times 8} \times 1$, or just $\frac{3 \times 6}{5 \times 8} = \frac{18}{40}$, much easier to work with than $\frac{504}{1,120}$. This is called **cancellation**—we can cancel the same factors in the numerator and denominator of a fraction.

But there is even more to cancellation. Can the fraction $\frac{18}{40}$ be reduced even further? Yes, if we write it as $\frac{9 \times 2}{20 \times 2}$, we can cancel the 2's and get $\frac{9}{20}$, which is the fraction **in lowest terms**. Lowest terms means it cannot be reduced any further. There are no numbers that divide evenly into both 9 and 20. This is also called **simplifying** the fraction.

A lot of cancellation can be done in your head by using your knowledge of the multiplication tables. For example, $\frac{24}{36}$ can be rewritten as $\frac{2}{3}$ by recognizing a common factor of 12 and canceling it from the top and bottom of the fraction. That's the important part of cancellation—the number must be a factor in the numerator *and* the denominator, and you can cancel it only once.

Now let's say we didn't recognize 12 as a common factor but instead knew that 6 was a common factor. We would get $\frac{24}{36} = \frac{4}{6}$. Then, since these are both even numbers, we should see that 2 is also a common factor, and we get $\frac{4}{6} = \frac{2}{3}$, the same result. So cancellation might take a few easy steps—but it's better than multiplying lots of numbers out unnecessarily.

Improper Fractions

How do you handle multiplication of mixed numbers, such as $2\frac{4}{5} \times 3\frac{2}{3}$? Fortunately, by using some basic facts that you probably "have always known," you can make multiplication of mixed fractions easier by changing them into **improper fractions** (the name comes from the fact that the numerator is larger than the denominator—they function the same as proper, or regular, fractions).

To change a mixed number to an improper fraction, we recognize that a mixed number, such as $2\frac{3}{4}$, actually has two parts, the whole number 2 plus the fractional part $\frac{3}{4}$. If we change the 2 to its equivalent in fourths, or $\frac{8}{4}$, then we see that $2\frac{3}{4} = \frac{8}{4} + \frac{3}{4} = \frac{11}{4}$.

This explanation assumes you know that, when adding fractions with the same denominators, you simply add the numerators, which we show in the section of this chapter on Addition and Subtraction of Fractions. The method for changing a mixed number to an improper fraction can be streamlined to three steps:

1. Multiply the whole number by the denominator.

2. Add the result to the numerator for the new numerator.

3. The denominator stays the same.

So for the numerator for the improper fraction for $2\frac{3}{4}$, we multiply 2×4 and then add it to 3. The denominator stays the same, and we get $\frac{11}{4}$ right away.

Often, the answer will yield another improper fraction, which we can change back to a mixed fraction. For example, $\frac{39}{5} = 7\frac{4}{5}$. How did we get that? Since a fraction is just another way to write division, we can use long division to get

$$
\begin{array}{r}
7 \\
5\overline{)39} \\
35 \\
\hline
4
\end{array}
$$

and the fractional part comes from the remainder of 4, which is the numerator of the fraction, and the denominator remains the same as the original one (it is the divisor).

Example 3.4.

Change the following mixed numbers to improper fractions:

a. $8\frac{2}{5} =$

b. $10\frac{2}{3} =$

c. $9\frac{3}{7} =$

d. $3\frac{7}{9} =$

e. $5\frac{3}{11} =$

f. $12\frac{1}{2} =$

Answer 3.4.

a. $8\dfrac{2}{5} = \dfrac{42}{5}$

b. $10\dfrac{2}{3} = \dfrac{32}{3}$

c. $9\dfrac{3}{7} = \dfrac{66}{7}$

d. $3\dfrac{7}{9} = \dfrac{34}{9}$

e. $5\dfrac{3}{11} = \dfrac{58}{11}$

f. $12\dfrac{1}{2} = \dfrac{25}{2}$

Example 3.5.

Write the following improper fractions as mixed numbers in lowest terms (use cancellation where possible):

a. $\dfrac{38}{5} =$

b. $\dfrac{37}{6} =$

c. $\dfrac{85}{9} =$

d. $\dfrac{52}{7} =$

e. $\dfrac{670}{80} =$

f. $\dfrac{85}{50} =$

Answer 3.5.

a. $\dfrac{38}{5} = 7\dfrac{3}{5}$

b. $\dfrac{37}{6} = 6\dfrac{1}{6}$

c. $\dfrac{85}{9} = 9\dfrac{4}{9}$

d. $\dfrac{52}{7} = 7\dfrac{3}{7}$

e. $\dfrac{670}{80} = \dfrac{67}{8} = 8\dfrac{3}{8}$

f. $\dfrac{85}{50} = \dfrac{17}{10} = 1\dfrac{7}{10}$, or $\dfrac{85}{50} = 1\dfrac{35}{50} = 1\dfrac{7}{10}$

So now we can multiply mixed numbers by changing them to improper fractions using cancellation where possible to simplify the calculation. Let's try a seemingly complicated multiplication to see how it can be simplified.

Let's find the product of $8\dfrac{2}{5} \times 3\dfrac{5}{6}$. Changing each of these numbers to an improper fraction, we have $\dfrac{42}{5} \times \dfrac{23}{6}$. Since 6 is a factor of the numerator 42 as well as the denominator 6, we can reduce this problem to $\dfrac{6 \times 7}{5} \times \dfrac{23}{6} = \dfrac{6 \times 7 \times 23}{6 \times 5} = 1 \times \dfrac{7 \times 23}{5} = \dfrac{161}{5} = 32\dfrac{1}{5}$. This one problem included all of the parts of multiplication of fractions discussed above. Even though this looks complicated because all the steps are written out here, usually the first two steps can be done in your head, with a little practice.

Let's try one more problem that looks even a little more difficult.

$$2\dfrac{8}{9} \times 3\dfrac{15}{16} \times \dfrac{6}{13} = \dfrac{26}{9} \times \dfrac{63}{16} \times \dfrac{6}{13} = \dfrac{1}{1} \times \dfrac{7}{4} \times \dfrac{3}{1} = \dfrac{21}{4} = 5\dfrac{1}{4}.$$

For this problem, we canceled 9's, 2's, and 13's from the numerators and denominators to reduce it to $\dfrac{1}{1} \times \dfrac{7}{4} \times \dfrac{3}{1}$, a much more manageable problem. We did the cancellation in two steps (first the 9's and 2's, and then the 13's), although all of the cancellations could have been done at the same time or in a different order.

Division of Fractions

Division of fractions follows the same rules as multiplication of fractions, but division is always done with only two fractions. Division of fractions is usually written with a $\div$ sign. If it takes the form of a fraction with a fractional numerator and a fractional denominator, such as $\dfrac{\frac{3}{4}}{\frac{1}{2}}$, rewrite it with the ÷ sign. Then all that needs to be done is to **invert** the divisor (second term) and multiply following the rules for multiplication of fractions from the last section. Inverting a fraction means simply reversing the numerator and denominator. The term for this is finding the **reciprocal**.

An example will show what is meant. Let's take our stacked fraction above, $\dfrac{\frac{3}{4}}{\frac{1}{2}}$, and first rewrite it as $\dfrac{3}{4} \div \dfrac{1}{2}$. Now we convert it into a multiplication problem by inverting the $\dfrac{1}{2}$, and our equivalent problem becomes $\dfrac{3}{4} \times \dfrac{2}{1}$. Recognizing that 2 is a factor of 4, we get $\dfrac{3}{2} \times \dfrac{1}{1} = \dfrac{3}{2} = 1\dfrac{1}{2}$.

When dividing mixed numbers, first change to improper fractions and then invert and multiply. For example, $2\dfrac{3}{4} \div \dfrac{5}{8} = \dfrac{11}{4} \times \dfrac{8}{5} = \dfrac{22}{5} = 4\dfrac{2}{5}$. Here, 4 is a factor of 8, so we canceled before doing the multiplication.

Example 3.6.

Perform the following divisions:

a. $\dfrac{3}{7} \div \dfrac{5}{6} =$

b. $2\dfrac{1}{4} \div \dfrac{3}{8} =$

c. $2\dfrac{2}{9} \div \dfrac{5}{12} =$

d. $3\dfrac{4}{7} \div \dfrac{5}{12} =$

e. $4\dfrac{11}{13} \div 2\dfrac{11}{26} =$

Answer 3.6.

a. $\dfrac{3}{7} \times \dfrac{6}{5} = \dfrac{18}{35}$

b. $\dfrac{9}{4} \times \dfrac{8}{3} = 6$

c. $\dfrac{20}{9} \times \dfrac{12}{5} = \dfrac{16}{3} = 5\dfrac{1}{3}$

d. $\dfrac{25}{7} \times \dfrac{12}{5} = \dfrac{60}{7} = 8\dfrac{4}{7}$

e. $\dfrac{63}{13} \div \dfrac{63}{26} = \dfrac{63}{13} \times \dfrac{26}{63} = 2$

Addition and Subtraction of Fractions

The most important rule about adding or subtracting fractions is that *the denominators of the terms must be the same*. Then you simply add or subtract the numerators and keep the denominators the same. For example, $\dfrac{1}{7} + \dfrac{4}{7} = \dfrac{5}{7}$ and $\dfrac{7}{5} - \dfrac{4}{5} = \dfrac{3}{5}$. Pretty straightforward, but what if the denominators are not the same? How do you get the denominators to be the same?

The answer to this is to find the **common denominator**, which is based on the facts that (1) any number divided by itself equals 1 and (2) multiplication by 1 doesn't change the value. For example, let's say we want to add $\dfrac{1}{7} + \dfrac{2}{5}$. If we multiply $\dfrac{1}{7}$ by $\dfrac{5}{5}$, we get $\dfrac{5}{35}$, which indeed is the same as $\dfrac{1}{7}$. Likewise, if we multiply $\dfrac{2}{5}$ by $\dfrac{7}{7}$, we get $\dfrac{14}{35}$, which is the same as $\dfrac{2}{5}$. So we have changed the original addition problem to one in which each term has the same denominator (35), and we just add the numerators to get $\dfrac{1}{7} + \dfrac{2}{5} = \dfrac{5}{35} + \dfrac{14}{35} = \dfrac{19}{35}$.

So we see that one way to get a common denominator is to multiply each term by the fraction (equal to 1) that is equal to the other denominator divided by itself. Then the new denominators are the same—just the product of the original two denominators. That works fine if the numbers are manageable, but what about adding something like $\dfrac{5}{27} + \dfrac{7}{54}$? Do we really want to multiply 27 by 54? Can we find another way to do this problem?

We can. We can find the *lowest* **common denominator** (LCD) to simplify our work. We know 27 × 54 will work as a common denominator, but it isn't the lowest one. If we recognize that 54 = 2 × 27, we can use 54 as our denominator, and the problem becomes $\dfrac{5}{27}\left(\dfrac{2}{2}\right) + \dfrac{7}{54} = \dfrac{10}{54} + \dfrac{7}{54} = \dfrac{17}{54}$. Much easier. Remember, we are looking for a number that all denominators can divide evenly into, and then we adjust each number to have that denominator.

It's not always straightforward to find the lowest common denominator. Sometimes it is obvious, such as for $\dfrac{5}{9} - \dfrac{2}{3}$, where the LCD is 9. This subtraction thus becomes $\dfrac{5}{9} - \dfrac{2}{3}\left(\dfrac{3}{3}\right) = \dfrac{5}{9} - \dfrac{6}{9} = -\dfrac{1}{9} = \dfrac{-1}{9}$.

Notice that the minus sign can appear before the fraction or in the numerator—it is the same thing. In fact, it can even appear in the denominator, although this isn't usual.

Sometimes it takes a little work to find the LCD, such as when adding $\frac{1}{4}+\frac{1}{5}+\frac{1}{6}$. What number do all three denominators divide evenly into? The best way to do this is to look at the largest denominator and its multiples (here it would be 6, 12, 18, 24, 30, 36, 42, 48, 54, 60, . . .) until you come to one that all the other denominators also divide evenly into. Here, we have to go all the way out to 60 to find the lowest common denominator, and the problem becomes

$$\frac{1}{4}+\frac{1}{5}+\frac{1}{6} = \frac{1}{4}\left(\frac{15}{15}\right)+\frac{1}{5}\left(\frac{12}{12}\right)+\frac{1}{6}\left(\frac{10}{10}\right) = \frac{15}{60}+\frac{12}{60}+\frac{10}{60} = \frac{37}{60}.$$

If you find a common denominator that isn't the lowest one, such as 120 for the problem given above, you still get the correct answer—it just takes a little longer:

$$\frac{1}{4}+\frac{1}{5}+\frac{1}{6} = \frac{1}{4}\left(\frac{30}{30}\right)+\frac{1}{5}\left(\frac{24}{24}\right)+\frac{1}{6}\left(\frac{20}{20}\right) = \frac{30}{120}+\frac{24}{120}+\frac{20}{120} = \frac{74}{120} = \frac{2\times37}{2\times60} = \frac{37}{60}.$$

HINT

When you add fractions, do not add the denominators—you will get the wrong answer. Usually, you can spot this because the answer doesn't seem right. For example, if you add $\frac{2}{3}+\frac{1}{2}$ and get $\frac{2}{5}$, you should recognize that something is wrong because $\frac{2}{5}$ is less than either of the fractions you added, and certainly their sum must be larger than either one of them.

To add mixed numbers, add the whole parts and then add the fractional parts. If the fractional parts add up to an improper fraction, change it to a mixed number, which then gets added to the whole number sum. For example, for $3\frac{7}{12}+2\frac{5}{6}$, the whole numbers add to $3+2=5$, and the fraction parts add as follows: $\frac{7}{12}+\frac{5}{6}=\frac{7}{12}+\frac{10}{12}=\frac{17}{12}=1\frac{5}{12}$. Add the 1 to the whole number 5, and the answer is $3\frac{7}{12}+2\frac{5}{6}=5+1\frac{5}{12}=6\frac{5}{12}$.

Example 3.7.

Lydia is making a bridal dress from a pattern that calls for $3\frac{1}{3}$ yards of fabric for the skirt, $1\frac{5}{6}$ yards for the bodice, and $4\frac{3}{4}$ yards for the train. Just to be sure she has enough fabric, she wants an additional yard to be added to her order.

a. How much fabric should she buy?

b. If the fabric is priced at $36 per yard, what is the total cost?

c. If Lydia wanted to round the purchase up to the next whole number of yards, how much more should she order?

d. How much more would this rounded-up amount cost?

Answer 3.7.

a. $10\frac{11}{12}$ yards. Add all the yardage that Lydia needs: $3\frac{1}{3} + 1\frac{5}{6} + 4\frac{3}{4} + 1$. (Don't forget the "additional yard" she wanted to add to the purchase.) Add the whole numbers to get $3 + 1 + 4 + 1 = 9$. Then add the fractional parts. The LCD for these fractions is 12, so the fractions become $\frac{4}{12} + \frac{10}{12} + \frac{9}{12} = \frac{23}{12} = 1\frac{11}{12}$. Adding the whole number total (9) and the fraction total ($1\frac{11}{12}$) gives $10\frac{11}{12}$.

b. $393. To find the cost, multiply the cost per yard by the yardage: $\$36 \times 10\frac{11}{12}$. This is the same as $\$36(10) + \$36(\frac{11}{12}) = \$360 + \$33 = \$393$.

c. 11 yards. The total purchase of $10\frac{11}{12}$ yards rounds up to 11 yards.

d. $3.00. The difference between 11 yards and $10\frac{11}{12}$ yards is $11 - 10\frac{11}{12} = \frac{1}{12}$ yard. At $36 per yard, this cost would be $\$36 \times \frac{1}{12} = \$3$. Another way to calculate this is to find the cost of 11 yards ($\$36 \times 11$), or $396, and subtract the cost of $10\frac{11}{12}$ yards from part (b): $\$396 - \$393 = \$3$.

Ratios and Proportions

A **ratio** is simply a way of comparing two numbers or expressing the relation between two quantities. A ratio can be expressed as a fraction or a decimal, and often is written with a colon between the numbers. For example, a small company employs 3 men and 7 women. Then we can say the ratio of men to women in that company is 3:7, or $\frac{3}{7}$, both read as "3 to 7" because we are stating a ratio. We can even say there is .4286 of a man for each 1 woman, by dividing 7 into 3, although this is a little awkward to visualize.

As another example, if the results of an election are a landslide in which one candidate received 5 times the number of votes of the other candidate, the ratio of votes is 1:5. However, the fractions of the votes received by the candidates are $\frac{1}{6}$ and $\frac{5}{6}$ (not $\frac{1}{5}$ for the losing candidate, as you might think). The ratio is 1:5, so the whole must be the sum, or $1 + 5 = 6$.

A **proportion** is the way to express that two ratios are equal. Again, the most common ways to write a proportion are with colons (3:7 = 6:14) or fractions ($\frac{3}{7} = \frac{6}{14}$), which is said as "3 is to 7 as 6 is to 14." We will see proportions again in Chapter 5, where we discuss solutions to algebra problems such as "If the ratio of men to women in the workplace is 3 to 7, how many men are in an office with 70 women?" Maybe you figured that out already. The ratio is $\frac{3}{7}$, which is the same as $\frac{30}{70}$, so for 70 women, there are 30 men. If you didn't see this right away, it's okay—this solution actually involves algebra.

Example 3.8.

Two-thirds of a Spanish class have never studied Spanish before. In a class of 30 students, what is the ratio of those who have studied Spanish to those who haven't?

Answer 3.8.

2:1. Be careful to keep the numbers straight in this problem. For the number who have never studied Spanish before, it is $\frac{2}{3} \times 30 = 20$. That means the remainder, $30 - 20 = 10$ students have studied Spanish before. The question asks for "the ratio of those who have studied Spanish to those who haven't," so the ratio has to follow that order: 20 to 10, or 2:1.

Percentages

The word *percent* comes from per hundred because "cent" refers to 100 (100 cents in a dollar, 100 years in a century, etc.). So a percentage is just a way to express a fraction with 100 as the denominator. In fact, the percent sign (%) came into use as a shorthand for $\frac{1}{100}$, with the two 0s in the % sign being the two 0s in the number 100. So to convert a percentage to a fraction, drop the percent sign and multiply by $\frac{1}{100}$. Therefore, 20% means $20 \times \frac{1}{100} = \frac{20}{100} = \frac{1}{5}$, and 100% means $100 \times \frac{1}{100} = \frac{100}{100} = 1$, the whole thing.

CALCULATOR

PERCENTAGES

The important keys for percentages are $\boxed{\text{2nd}}\,\boxed{(}$ (which gives "% of"), the toggle key $\boxed{<>}$, and $\boxed{\text{enter}}$, scrolling $\boxed{\updownarrow}$, plus any of the basic operation keys.

For example, to find 12% of 135, enter 12, $\boxed{\text{2nd}}\,\boxed{(}$, 135 and press $\boxed{\text{enter}}$. The screen will look like 12%135 and will display the answer 16.2. Use the toggle key $\boxed{<>}$ to change the decimal to a fraction.

To change a number to a percentage, use the $\boxed{\text{2nd}}\,\boxed{)}$ combination. For example, to find the decimal .00236 expressed as a percentage, enter .00236 $\boxed{\text{2nd}}\,\boxed{)}$ and press $\boxed{\text{enter}}$. The result will show as 0.236%.

To display a fraction as a percentage, enter the fraction with the $\boxed{\frac{n}{d}}$ key, scroll right $\boxed{\rightarrow}$, then $\boxed{\text{2nd}}\,\boxed{)}$ and press $\boxed{\text{enter}}$. For example, for $\frac{3}{4}$, press 3, then $\boxed{\frac{n}{d}}$, then 4, scroll right $\boxed{\rightarrow}$, $\boxed{\text{2nd}}\,\boxed{)}$, and press $\boxed{\text{enter}}$. The result looks like $\frac{3}{4}$ ›%, and the answer is 75%. This function is not on the GED® Calculator Reference pull-down, but it could come in handy.

The Relations Among Decimals, Fractions, and Percentages

Fractions and decimals are closely related, as are decimals and percentages. You should know how to convert from one form to another. Fortunately, the calculator provided on the GED® test does the conversion from decimal to fraction and back just by pressing (enter) and then the toggle (<>) key.

Changing a Fraction to a Decimal

To change a fraction to a decimal, simply divide the numerator by the denominator. Since both the numerator and denominator are whole numbers (which means the decimal point, even if not written, is at the end of the number), the decimal point in the answer is simply above the decimal point in the dividend. The division proceeds as division of decimals did. As an example, $\frac{3}{4}$ becomes .75 as follows:

$$
\begin{array}{r}
.75 \\
4\overline{)3.00} \\
\underline{28} \\
20 \\
\underline{20}
\end{array}
$$

Not all fractions will have decimals that end so neatly. In fact, most will not. The GED® test works only with **rational numbers**, which are defined as "any number that can be written as a fraction." Rational numbers include all whole numbers since every whole number can be written as a fraction with 1 in the denominator.

You should be familiar with some fractions, and they probably are not new to you. From our money system, you know that $\frac{1}{4}$ (a quarter) is .25, $\frac{2}{4} = \frac{1}{2}$ (2 quarters, or a half) is .50, and $\frac{3}{4}$ (3 quarters) is .75. The fractions involving thirds should be familiar to you: $\frac{1}{3} = .3\overline{3}$ and its double, $\frac{2}{3} = .6\overline{6}$. The lines over the last two digits mean those numbers are repeated forever.

Other interesting fractions include all of the fifths, which have a decimal that is just double the numerator (for example, $\frac{1}{5} = .20$, $\frac{2}{5} = .40$, $\frac{3}{5} = .60$, and $\frac{4}{5} = .80$), and the ninths, which just repeat the numerator forever (for example, $\frac{1}{9} = .1\overline{1}$, $\frac{2}{9} = .2\overline{2}$, $\frac{4}{9} = .4\overline{4}$, and $\frac{5}{9} = .5\overline{5}$). You needn't memorize fractions because you can always get them by dividing the numerator by the denominator as we did at the beginning of this section, but by knowing some by sight, you won't waste time during the test.

Changing a Decimal to a Fraction

The placement of the digits in a decimal gives them an automatic denominator for conversion to fractions. Recall that the placeholders for decimals, reading from the decimal point to the right, are tenths, hundredths, thousandths, ten thousandths, etc. Also, a decimal is read as if the digits were a whole number, which becomes the numerator, and the rightmost digit names the placeholder, which is the denominator. This means that .75 is read as "seventy-five hundredths," which can be written as $\dfrac{75}{100} = \dfrac{3}{4}$ (after dividing both numerator and denominator by 25). Likewise, .036 is 36 thousandths, or $\dfrac{36}{1000} = \dfrac{9}{250}$ (recognizing that 4 is a common factor). A mixed number involving a decimal can be converted into a mixed number involving a fraction or can be an improper number, such as $2.7 = 2\dfrac{7}{10} = \dfrac{27}{10}$. Note, we still use the rightmost placeholder, which is tenths here, as the denominator.

Percentages and Decimals

Converting percentages to decimals just involves dropping the percent sign and moving the decimal point in the percentage two places to the left: two places because hundredths has two placeholders, to the left because the decimal equivalent is 100 times smaller than the percentage number. Confusion may arise when you have to fill in "phantom" places with zeros, such as in 3%, which equals .03, or when the percentage is itself a decimal, such as 1.2%, which is .012. The rule doesn't change—move the decimal point two places to the left when converting percentages to decimals.

In contrast, when converting decimals to percentages, the decimal movement is to the right. Again, if you remember that .50 is 50%, you shouldn't get confused about which direction to move the decimal point. For example, .23 is 23%, .675 is 67.5%, and .04 is 4%.

HINT

Sometimes it is confusing whether to move the decimal point to the right or the left when converting from percentages to decimals, so just keep this equivalence in mind:

50% is .50, or one-half.

This will help you remember that the decimal in 50% is moved two places to the left. Likewise, the decimal in .50 is moved two places to the right when changing .50 to 50%.

Real-World Situations

Interest

The formula $I = prt$, where I is interest, p is principal (or amount invested), r is (annual) rate, and t is time (in years), is given on the pull-down formula sheet on the GED® test. What you have to remember is that, to do the calculation, the rate has to be converted to a decimal. So a 5% interest rate means to multiply by .05, not 5. For example, to calculate the interest on a $5,000 loan at 6% for 12 months, you have to multiply $5,000 × .06 × 1, to get $300.

In the above calculation, the 1 represents time (t) because this equation uses years, so 12 *months* has to be converted to 1 *year*. To convert months to years, divide the number of months by 12 because there are 12 months in a year. Then 3 months would be entered in an interest equation as $\frac{3}{12}$, or $\frac{1}{4}$ of a year.

Markups and Markdowns

The general idea of retail is that the shop owner buys an item (at **cost**) and applies a **markup** to cover expenses and profit. This determines the **price** to the customer.

If an item isn't selling as well as it should, the shop owner offers a **markdown** (or **discount**) to entice the shopper to buy the item. Sometimes there are multiple discounts. The difference between the original price and the discount is the discounted price, called the **sale price**. This section gives a summary of how to do problems that involve markups and discounts.

Usually, the rate of markup or discount is given as a percentage, but when doing the calculation, convert percentages to decimals.

To determine the price, multiply the cost (wholesale) by the markup percentage to get the markup and add it to the cost. So if an item cost the shop owner $100 and he wants a profit of 40%, the markup will be ($100) (.40) = $40, and the price will then be $100 + $40 = $140. So, in general, for markups,

$$\text{Markup} = \text{Cost} \times \text{Percentage (in decimals)}$$
$$\text{Price} = \text{Cost} + \text{Markup}$$

To determine the *sale* price, the general formula is likewise

$$\text{Discount} = \text{Price} \times \text{Percentage (in decimals)}$$
$$\text{Sale Price} = \text{Price} - \text{Discount}$$

So the idea is the same, except a markup *adds* the percentage and a discount *subtracts* the percentage. For example, Mr. Hoffman paid a wholesale price of $7.50 (his cost) for an item and he wants to make a 50% profit. His markup is $7.50 × .50 = $3.75, so the price to the customer is $7.50 + $3.75 = $11.25. The markup is added to the cost to the shop owner.

Now, let's look at another example. Robin really wants to buy a particular dress, but at $100, it is beyond her budget. Mrs. Marquez, the owner of the dress shop, is expecting a new shipment of dresses, so she wants to sell more of what she already has on the rack. Therefore, she offers a 40% discount on a rack of dresses, including the one Robin likes. Now the dress will cost $100 × .40 = $40 less, or $100 − $40 = $60. (Robin is happy.) The discount is subtracted from the price of the dress.

HINT

A **markup** is *added* to the *cost*, and a **discount** is *subtracted* from the *price*. That makes sense—a shop owner wouldn't sell something at less than it cost, and a shopper wouldn't buy an item that cost more than the original price.

Now we just need a way to get the percentage if we know the cost and price, or the original price and the discounted price. The general statement for discounts (and markups) is

"The discount (or markup) is a percentage of the price (or cost)."

Said in algebra terms, where "is" means "equals" and "of" means "multiply," as shown in Chapter 5, we get

$$\text{Discount} = \text{Price} \times \text{Percentage}$$

or

$$\text{Markup} = \text{Cost} \times \text{Percentage},$$

which we saw already, but now it is a simple equation that can work with all percentages. Then the percentage (or discount or markup) is given by

$$\text{Percentage} = \frac{\text{change}}{\text{original}},$$

where "change" means the *increase* over the original cost for the shop owner or the *decrease* from the original price for the shopper, and "original" means either the original cost or the original price (before the markup or markdown).

Watch out for double discounts. At the end of the season, many stores discount already discounted items. For example, a 50% discount may be followed by a 40% discount. But this is *not* the same as a 90% discount. That is because the 40% discount is a new discount, and the "original" price it is based on is the former discounted price. For example, if an item cost $60 and is discounted 50%, it now cost $30. When the store offers an additional 40%, it is off the $30 price, not the original price, so it is $12. The item will now be priced at $18 (still a good bargain).

Let's look at some examples to show how markups and markdowns work.

Example 3.9.

The owner of a dress shop buys a fancy dress for $50.

    a. She thinks her customers will really like it, so she marks it up 90%. What is the price for the dress?

    b. She judged that all wrong, so she offers it for sale at a whopping 90% off. Will she break even?

a. $95. Price = Cost + (Cost × Markup Percentage), so the dress will be priced at $50 + ($50 × .90) = $50 + $45 = $95.

b. No. Now the dress is offered at discount of 90%. So the Sale Price = Price − (Price × Discount). Therefore, the dress finally sells for $95 − ($95 × .90) = $95 − $85.50 = $9.50, a lot less than the original cost of $50.

Example 3.10.

For the scenario in Example 3.9, what percentage discount should the shop owner use so that she does break even?

Answer 3.10.

47%. The original price is $95 and the shop owner just wants to get the cost of $50 back, so the discounted price will be $50. So the percentage discount should be calculated from

$$\text{Percentage} = \frac{\text{change}}{\text{original}},$$

where the change will be $45.00 (the original price of $95 minus the new price of $50) and the original price is $95.00. Then the percentage discount is $\frac{45}{95}$ = 47% off.

In reality, this is such a strange number that the shop owner would probably market the dress at "45% off" or even "40% off" to get a little profit. If the discount were rounded up to 50% off, the shop owner would lose a little money. Can you figure how much that loss would be? [Fifty percent off the price of $95 is ($95) (.50) = $47.50, so since the cost to the shop owner was $50, that would be a $2.50 loss.]

Exercises

1. Jeremy is taking a bike trip. He wants to go 130 miles in 8 hours. If he rides at a steady speed, he will ride [ ] miles each hour.

2. The fraction $\frac{3}{7}$ is $\left\{ \begin{array}{l} \text{greater than} \\ \text{equal to} \\ \text{less than} \end{array} \right\}$ the fraction $\frac{6}{14}$.

3. Alexis is planning to make hot chocolate from a mix for her daughter and four of her friends. Each cup of hot chocolate requires a half cup of mix. How many cups of mix will Alexis need?

    A. 2 cups

    B. $2\frac{1}{2}$ cups

    C. 4 cups

    D. 5 cups

4. Victor bought $\frac{3}{4}$ pound of shrimp that were advertised as being $45-50$ shrimp to the pound.

    a. Approximately how many shrimp did Victor buy?

        A. 28

        B. 35

        C. 40

        D. 45

    b. Victor wants to make shrimp cocktail with 8 shrimp in each serving. How many servings can he make?

        A. 8

        B. 6

        C. 4

        D. 2

5. *Do not use your calculator for this exercise.* An $8\frac{1}{2}$-ounce bag of potato chips says there are 9 chips per serving and that there are "about 8" servings in the bag.

    a. Approximately ☐ chips are in the bag.

    b. One serving is about ☐ ounces to the nearest whole number.

6. A recipe calls for twice as much sugar as flour. If we use 3 cups of sugar, how much flour do we use?

    A. 6 cups

    B. 1 cup

    C. 3 cups

    D. $1\frac{1}{2}$ cups

7. Carlos works part-time for $9.00 per hour. At the beginning of the year, he will get a 10% raise. He will then make $ [＿＿＿＿] per hour.

8. A movie theater is open from noon to 11:30 p.m. If a complete showing of a certain movie, including trailers, takes 2 hours and 15 minutes, how many times in a day can the cinema run a complete showing of that movie? (Hint: Every 15 minutes is $\frac{15}{60} = \frac{1}{4}$ hour, and every 30 minutes is $\frac{30}{60} = \frac{1}{2}$ hour.)

    A. 5

    B. $5\frac{1}{9}$

    C. 6

    D. Cannot tell from the information given.

9. The budget for a small office allows $\frac{1}{4}$ of its gross income for rent and utilities and $\frac{1}{3}$ of the rest for salaries. The fraction of the gross income budgeted for salaries is $\begin{Bmatrix} \text{more than} \\ \text{equal to} \\ \text{less than} \end{Bmatrix}$ the fraction budgeted for rent.

10. Lou can do $\frac{1}{2}$ of a job by himself in one day. Joe can do $\frac{3}{8}$ of the job by himself in one day. If they work together, how much of the job is left undone?

    A. $\frac{1}{2}$

    B. $\frac{3}{8}$

    C. $\frac{7}{8}$

    D. $\frac{1}{8}$

11. Sarah bought 40 half-gallon plastic containers for her kitchen. She made 5 gallons of soup and wants to freeze it.

   a. Sarah will need ⬚ half-gallon containers.

   b. If Sarah fills each container only $\frac{2}{3}$ full to allow for expansion during freezing, how many more containers will she need?

      A. 15

      B. 10

      C. 5

      D. 0

12. Reduce $\frac{26}{286}$ to lowest terms.

      A. $\frac{13}{143}$

      B. $\frac{1}{11}$

      C. $\frac{2}{22}$

      D. $\frac{26}{286}$

13. Greg got a bonus of $500 for a project he completed before deadline. He wants to put his money in a checking account that pays 1.2% annual interest. If he keeps the $500 in the bank for 18 months, how much money will he have earned?

      A. $9.00

      B. $90.00

      C. $108.00

      D. $6.00

14. A bicycle shop can get a popular bike that normally sells for $375 for a cost of only $150. What percent markup does this represent?

     A. 250%

     B. 150%

     C. 225%

     D. 167%

Solutions

Answer 1. $16\frac{1}{4}$. He would ride 130 miles in 8 hours, or $\frac{130}{8} = 16\frac{1}{4}$ miles each hour.

Answer 2. Equal to. $\frac{3}{7} = \frac{6}{14}$. Multiply $\frac{3}{7}$ by $\frac{2}{2}$.

Answer 3. (B) $2\frac{1}{2}$. First, there are going to be 5 cups of hot chocolate (not 4—you have to include Alexis's daughter), each needing $\frac{1}{2}$ cup of mix. So Alexis needs $5 \times \frac{1}{2} = 2\frac{1}{2}$ cups of mix.

Answer 4. **a.** (B) 35. Three-quarters of 45 to 50 shrimp give a range of 33.75 to 37.50 shrimp. Since there shouldn't be a fraction of a shrimp, there are 33 to 37 shrimp in his purchase.

b. (C) 4. Victor needs 32 shrimp to make 4 servings. More than that would require 40 shrimp. Even if you didn't get part (a) correct, you still can figure out that is the only logical answer since 2 servings would be only 16 shrimp, and he clearly has more than that, and 6 or 8 servings would take a pound or more of the shrimp.

Answer 5. **a.** 72. About 8 servings at 9 chips per serving means there are $8 \times 9 = 72$ chips.

b. 1. About 8 servings from an 8.5-ounce bag means each serving is $8\frac{1}{2} \div 8$, or about 1 ounce.

Answer 6. (D) $1\frac{1}{2}$ cups. Twice as much sugar as flour means half as much flour as sugar. So if we use 3 cups of sugar, we use $3 \times \frac{1}{2} = \frac{3}{2} = 1\frac{1}{2}$ cups of flour.

Answer 7.   9.90. A 10% raise means a raise of $9.00 × .10 = $.90. Added to Carlos's present hourly rate, that will be $9.00 + $.90 = $9.90. The dollar sign is already given, so state your answer without a dollar sign.

Answer 8.   The cinema is open for $11\frac{1}{2}$ hours a day. The movie takes $2\frac{1}{4}$ hours. So the number of times the movie can be shown is $11\frac{1}{2} \div 2\frac{1}{4}$, or $\frac{23}{2} \div \frac{9}{4}$, which is $\frac{23}{2} \times \frac{4}{9}$ (when dividing fractions, invert the divisor and multiply). Enter any of these computations on your calculator to get $\frac{46}{9}$, which is $5\frac{1}{9}$. Since we are asked for complete showings, the answer is 5.

Answer 9.   Equal to. You don't have to know what the gross income is to solve this problem. After $\frac{1}{4}$ of the income is used for rent and utilities, $\frac{3}{4}$ is left, the problem says $\frac{1}{3}$ of that goes to salaries. So $\frac{1}{3} \times \frac{3}{4} = \frac{1}{4}$ of the income goes to salaries, the same as that for rent.

Answer 10.   (D) $\frac{1}{8}$. Answer choices (A) and (B) are incorrect because they are how much each man can do alone, but they are working together. They can do $\frac{1}{2} + \frac{3}{8} = \frac{7}{8}$ of the job by working together. But the question is what fraction of the job is *left undone*, so it is the remaining $\frac{1}{8}$. Read carefully!

Answer 11.   **a.** 10. The number of gallons Sarah bought doesn't make a difference, as long as she bought enough. The important numbers in this problem are that Sarah wants to put 5 gallons of soup into half-gallon containers. So she divides the 5 gallons by $\frac{1}{2}$, so $5 \div \frac{1}{2}$ = 5 × 2 = 10. Sarah will need 10 containers.

b. (C) 5. Now each container holds only $\frac{2}{3}$ of its capacity, or $\frac{2}{3} \times \frac{1}{2} = \frac{1}{3}$ gallon per container. So now the calculation for how many containers she will need is $5 \div \frac{1}{3}$ = 5 × 3 = 15. But the question is how many *more* containers she will need, so the answer is 15 − 10 = 5 more. Be sure to read each problem to see what it really is asking.

Answer 12. (B) $\frac{1}{11}$. To reduce a fraction to lowest terms means dividing the numerator and denominator by the greatest common factor of both. If you didn't recognize that 26 is the greatest common factor, at least 2 is a factor since both numbers are even. Dividing the numerator and denominator by 2 gives $\frac{13}{143}$, and since 13 is prime, the only choice is to see whether 13 divides evenly into 143, which it does ($\times$ 11). Dividing numerator and denominator by 13 gives $\frac{1}{11}$.

Answer 13. (A) $9.00. The formula for interest is $I = prt$ (this is on the GED® test formula sheet). The calculation is thus $500 $\times$ .012 $\times$ 1.5 = $9.00. Remember that 1.2% is .012 in decimal notation, which is what is required for this formula, and 18 months $= \frac{18}{12}$ years $= \frac{3}{2}$ years = 1.5 years.

Answer 14. (B) 150%. The equation is Percentage $= \dfrac{\text{Change}}{\text{Original}}$, where Change is the difference between the price and the cost, and Original refers to the cost. So the calculation is $\dfrac{375-150}{150} = \dfrac{225}{150} = 1.5$, which is 150%. Just to check this, the shop owner's profit will be 150% of his $150 cost, or $225. So he sells the bike for his cost plus profit, or $150 + $225 = $375.

Power Play

Powers

Powers have to do with multiplication. They are a shorthand for repeated factors. But what does that mean? If you multiply 2×2, you get 4. If you multiply $2 \times 2 \times 2$, the product is 8, and if you multiply $2 \times 2 \times 2 \times 2$, the answer is 16. But shouldn't there be a shorter way to indicate the multiplication of a string of, say, eight 2's than by writing $2 \times 2 \times 2 \times 2 \times 2 \times 2 \times 2 \times 2$?

This is where exponents come in. They indicate the **power**, or the number of repeated factors. We can write $2 \times 2 \times 2 \times 2 \times 2 \times 2 \times 2 \times 2$ as 2^8, which we say is "2 to the eighth power." The 8 is an **exponent** and the 2 is called a **base**. Notice that 2^8 means there are eight 2's multiplied together. This is not the same as 2 multiplied by itself eight times—if you count, there are only seven "$\times$" signs.

Two particular powers are quite important: 2 and 3. An exponent of 2 is almost always called **squared** rather than "to the second power," and an exponent of 3 is almost always called **cubed** rather than "to the third power." As you might guess by these words, they derive from the geometry of squares and cubes, as we will see in Chapter 7. For now, just remember that $5^2 = 25$ because it is 5×5, and $2^3 = 8$ because it is $2 \times 2 \times 2$. Exponents are used a lot in algebra, the topic of Chapters 5 and 6, and we will discuss them in more detail there.

The powers of 0 and 1 are special. Anything to the zero power equals 1. *Anything*. So $10^0 = 1$ and $524^0 = 1$ also. Anything to the first power is itself. That makes sense from the definition of power. So $10,524^1 = 10,524$, and $0^1 = 0$. (Note that 0^0 is undefined because it cannot be 1 and 0 at the same time.)

CALCULATOR

POWERS

The important keys for calculating **powers** are the x^2 and $\wedge$ keys, the toggle key $\Longleftrightarrow$, and enter.

For example, to find 3^2, enter 3, then the x^2 key, and enter. The screen will show 3^2 and give 9 as the answer. To find 4^3, enter 4, then the $\wedge$ key, the 3 the exponent, and enter. The screen will show 4^3 and give 64 as the answer. It seems like all that is needed is the $\wedge$ key because it does work for squares, but squares are so common, they get a "one-hit" key of their own. If you enter a decimal number, such as 2.5, its square is not a whole number (it is 6.25), but as for all decimal numbers, the toggle key $\Longleftrightarrow$ will convert it to a fraction.

Even though the calculator is handy for exponents, it is a good idea to know the powers of 2 at least up to the sixth power, 3 up to the third power, and the squares of all the numbers from 1 to 10, and not have to spend the time entering them on the calculator (although that is always an option if you forget):

$2^2 = 4 \qquad 2^3 = 8 \qquad 2^4 = 16 \qquad 2^5 = 32 \qquad 2^6 = 64 \qquad 3^2 = 9 \qquad 3^3 = 27$

$4^2 = 16 \qquad 5^2 = 25 \qquad 6^2 = 36 \qquad 7^2 = 49 \qquad 8^2 = 64 \qquad 9^2 = 81 \qquad 10^2 = 100$

Negative Exponents

Exponents can also be negative, which means **reciprocal**. If you multiply a number by its reciprocal, you get 1. So if $2^3 = 8$, $2^{-3} = \dfrac{1}{8}$. Flip the value over the fraction bar, so to speak. If the negative exponent appears in the denominator of a fraction, then it flips up to the numerator. Therefore, $\dfrac{1}{3^{-2}} = 9$ (the reciprocal of $\dfrac{1}{9}$).

HINT

A negative exponent means nothing more than reciprocal—it has nothing to do with negative numbers.

Operations with Exponents

For addition or subtraction of numbers with the same base and exponent, simply add or subtract the coefficients and the base stays the same. For example, $2^3 + 6(2^3) - 2^4 = 7(2^3) - 2^4 = 56 - 16 = 40$. We combined the two terms with 2^3 to get $7(2^3)$, but the 2^3 and 2^4 terms cannot be combined similarly. We also could have just evaluated each term separately, $8 + 6(8) - 16 = 8 + 48 - 16 = 72$.

Multiplication and division of numbers with the same base, however, allows some streamlining—you don't have to evaluate each factor.

- For multiplication of numbers with the same base, add their exponents.

- For division of numbers with the same base, subtract their exponents (numerator exponent minus denominator exponent, or top exponent minus bottom exponent).

With the subtraction of exponents used for division, it is possible to get the negative exponents mentioned above. Some examples will show these rules in action. They are used extensively in algebra, as shown in Chapters 5 and 6.

Example 4.1.

Evaluate $2^3 + 2^3(2^2) - \dfrac{3^4}{3^5} + \dfrac{2^5}{2^3}$.

Answer 4.1.

$43\dfrac{2}{3}$. By using the rules for multiplication and division of exponents, this simplifies to $2^3 + 2^3(2^2) - \dfrac{3^4}{3^5} + \dfrac{2^5}{2^3} = 2^3 + 2^5 - 3^{-1} + 2^2 = 8 + 32 - \dfrac{1}{3} + 4 = 43\dfrac{2}{3}$. By brute force, this would have been $8 + 8(4) - \dfrac{81}{243} + \dfrac{32}{8}$, which eventually would have equaled $43\dfrac{2}{3}$.

Example 4.2.

Evaluate $\dfrac{6^{-4} \times 3^{12} \times 6^{-3}}{3^{24} \times 6^{-8} \times 3^{-11}}$.

Answer 4.2.

2. Since only multiplication and division are involved, we have several ways to evaluate this expression. The easiest is to put all of the factors with a base of 6 together and all of the factors with a base of 3 together and use the exponent rules. First, the 3^{12} in the numerator can flip down to the other base 3 factors in the denominator, where it becomes 3^{-12}. Likewise, let's flip 6^{-8} up with the other base 6 factors, where it will change to 6^8. Then we have $\dfrac{6^{-4} \times 6^8 \times 6^{-3}}{3^{24} \times 3^{-12} \times 3^{-11}}$, and when we add the exponents in the numerator and all the exponents in the denominator, we are left with $\dfrac{6}{3} = 2$. This is only one way to do the math—there are several others, and they will all equal 2.

Raising Powers to Powers

What happens when we raise an expression with a power to another power, such as $(2^3)^2$? We know that this answer should be $(8)^2 = 64$. But $64 = 2^6$. So what have we done with the exponents? When a power is raised to a power, we multiply the exponents. There is a difference between $(2^3)^2$ and $2^3(2^2) = 8(4) = 32 = 2^5$, which is multiplying two numbers with the same bases. In fact, these are the very examples that remind us what to do in each case.

The following rules apply to numbers *with the same base*:

1. When multiplying two numbers with the same base, add their exponents and keep the base the same.

2. When dividing two numbers with the same base, subtract their exponents and keep the base the same.

3. When raising a base number with an exponent to a power, multiply the exponents and keep the base the same.

HINT

To remember the above rules, just use $(2^3 \times 2^2)$, $\dfrac{2^3}{2^2}$, and $(2^3)^2$ because you can quickly construct the rules just by knowing the first six powers of two: 2, 4, 8, 16, 32, 64.

1. Rule 1: $(2^3 \times 2^2) = 8 \times 4 = 32 = 2^5$, and $5 = 3 + 2$, so when multiplying, add the exponents.

2. Rule 2: $\dfrac{2^3}{2^2} = \dfrac{8}{4} = 2 = 2^1$, and $1 = 3 - 2$, so when dividing, subtract the exponents.

3. Rule 3: $(2^3)^2 = (8)^2 = 64 = 2^6$, and $3 \times 2 = 6$, so when raising a power to a power, multiply the exponents.

Roots

Roots can be thought of as the opposite of powers. If $3^4 = 81$, we say, "3 to the fourth power is 81." The root sentence for this would be "The fourth root of 81 is 3," and in symbols it is written as $\sqrt[4]{81} = 3$. So this "root" is actually answering the question, "What number repeated as a factor four times gives us 81?"

The symbol for root is $\sqrt{}$, called a **radical**, where the number for which we are finding a root (X in the radical below), called the **radicand**, goes under the radical sign and the root we are finding (n in the radical below), called the **index**, is inserted as a smaller (in size) number at the left of the radical sign.

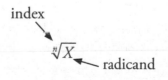

index

$\sqrt[n]{X}$ radicand

Because the most common index is 2, we don't even bother to write the little 2 on the radical. We just know that $\sqrt{}$ has index 2, and it is called a **square root**. If the index is 3, we do write the little 3, so we have $\sqrt[3]{}$, which has the special name of **cube root**. All of the other roots are called by their indexes: "fourth root," "fifth root," and so forth.

Fractional Exponents

Fractional exponents imply roots. The numerator still indicates power, but the denominator of the fractional exponent indicates root. For example, $8^{\frac{2}{3}}$ means either $\sqrt[3]{8^2}$ (the cube root of 8^2) or $\left(\sqrt[3]{8}\right)^2$, the square of the cube root of 8. It makes no difference whether you do the root or the power first. In this case, since the cube root of 8 is 2, it is easier to do the root first, and the answer is $8^{\frac{2}{3}} = 2^2 = 4$. Doing it the other way, we would have to remember $\sqrt[3]{64}$, which, of course, also gives us 4.

Roots of Negative Numbers

You can find square roots of positive numbers only. The square root of a negative number is called imaginary because there are no two identical numbers that multiplied together will produce a negative number (two positives multiplied together or two negatives multiplied together will always produce a positive result). However, for odd roots, negative radicands are fine because three negatives multiplied together will produce a negative number—for example, $(-3) \times (-3) \times (-3) = -27$, so $\sqrt[3]{-27} = -3$.

CALCULATOR

ROOTS

The important keys for calculating **roots** are the (2nd) (x²) keys (for square roots) and the (2nd) (∧) keys for other roots, the toggle key (<>), and (enter).

For square roots, enter the function first and then the number. For example, to find $\sqrt{25}$, enter the (2nd) (x²) keys, 25, and (enter). The screen will show $\sqrt{25}$ and 5 as an answer.

If the root is not a whole number, such as $\sqrt{8}$, the answer will come up in radical form, but the toggle key (<>) will convert it to a decimal. So for $\sqrt{8}$, the screen looks like $\sqrt{8}$ and the answer is $2\sqrt{2}$, but the toggle key converts this to 2.828427.

For cube roots and higher, enter the root number before the function keys. For example, to find $\sqrt[3]{27}$, enter 3 (the root), then the (2nd) (∧) keys, then 27 and (enter). It seems like all that is needed is the (2nd) (∧) keys, and it does work for square roots, but square roots are so common, they get a key of their own.

The counterparts of the powers you should know are the following roots (again, if you forget them, you can use the calculator):

$$\sqrt{4} = 2 \qquad \sqrt[3]{8} = 2 \qquad \sqrt[4]{16} = 2 \qquad \sqrt[5]{32} = 2 \qquad \sqrt[6]{64} = 2 \qquad \sqrt{9} = 3 \qquad \sqrt[3]{27} = 3$$

$$\sqrt{16} = 4 \qquad \sqrt{25} = 5 \qquad \sqrt{36} = 6 \qquad \sqrt{49} = 7 \qquad \sqrt{64} = 8 \qquad \sqrt{81} = 9 \qquad \sqrt{100} = 10$$

Simplifying Roots

Radicands can sometimes be factored, which makes finding the root easier. Take, for example, $\sqrt{36}$. If we factor the 36 into 4×9, we have $\sqrt{36} = \sqrt{4 \times 9} = \sqrt{4} \times \sqrt{9} = 2 \times 3 = 6$. Some seemingly difficult roots can be simplified by using this method. For example, $\sqrt{200} = \sqrt{2 \times 100} = \sqrt{2} \times \sqrt{100} = \sqrt{2} \times 10$, which is written as $10\sqrt{2}$.

The same simplification works when we have multiplications such as $3\sqrt{2} \times 4\sqrt{5}$. This is equivalent to $(3 \times 4)(\sqrt{2} \times \sqrt{5}) = 12\sqrt{10}$.

Likewise, if the radicand can be written as the division of two numbers, we can break it up into a fraction of roots, which we can divide (provided the roots have the same index). For example, $\frac{8\sqrt{15}}{4\sqrt{5}} = \frac{8}{4}\frac{\sqrt{15}}{\sqrt{5}} = 2\sqrt{3}$.

Just remember that we cannot combine sums, such as $3\sqrt{2} + 4\sqrt{5}$. Also, if the radicands are different, we cannot combine them. However, if we have $3\sqrt{2} + 5\sqrt{32}$, we can simplify the $5\sqrt{32}$ to $5\sqrt{16} \times \sqrt{2} = 5 \times 4 \times \sqrt{2} = 20\sqrt{2}$, and now we have

$$3\sqrt{2} + 5\sqrt{32} = 3\sqrt{2} + 20\sqrt{2} = 23\sqrt{2}$$

because we can combine like radicands.

Example 4.3.

Simplify $\sqrt{(4 + 28)}$.

Answer 4.3.

$4\sqrt{2}$. Remembering "do parentheses first," we rewrite it as $\sqrt{(4 + 28)} = \sqrt{32} = \sqrt{16 \times 2} = \sqrt{16} \times \sqrt{2} = 4\sqrt{2}$. Note that $\sqrt{(4 + 28)} \neq \sqrt{4} + \sqrt{28}$. The simplification applies only to products, not sums.

Example 4.4.

Simplify $\dfrac{\sqrt{32}}{\sqrt{2}}$.

Answer 4.4.

We see that $\dfrac{\sqrt{32}}{\sqrt{2}} = \sqrt{\dfrac{32}{2}} = \sqrt{16} = 4$.

Example 4.5.

Simplify $\sqrt{5} + \sqrt{125}$.

Answer 4.5.

$\sqrt{5} + \sqrt{125} = \sqrt{5} + \sqrt{25 \times 5} = \sqrt{5} + \sqrt{25} \times \sqrt{5} = \sqrt{5} + 5\sqrt{5} = 6\sqrt{5}$.

Order of Operations

Now that we have discussed all of the operations we will find in the rest of the book, let's make sure we know that there is an order to be followed when simplifying expressions.

1. **Parentheses** say "do me first," as was mentioned several times already. In other words, get rid of all of the parentheses ( ) or brackets [ ] or braces { }, working from the inside out by evaluating what is in them until they are all gone.

2. Evaluate any part of the expression that contains **exponents** next.

3. Then do any **multiplication** and/or **division** in order from left to right.

4. Last, do any **addition** and/or **subtraction** in order from left to right.

The mnemonic (a "fake" word whose first letters help memory) **PEMDAS** helps to remember the order:

Parentheses, **E**xponents, **M**ultiplication, **D**ivision, **A**ddition, and **S**ubtraction.

The multiplication and division should go in order from left to right and then the addition and subtraction should go in order from left to right as well.

"**P**lease **E**xcuse **M**y **D**ear **A**unt **S**ally" works as a reminder for some students.

Now let's see how order of operations works:

Example 4.6.

Evaluate $3^2 + \dfrac{16}{8} - (7 + 2^2) \times (5 - 3)^3$.

Answer 4.6.

-77. First we look at the parentheses. We can replace the first parentheses by 11 and the second parentheses by 2:

$$3^2 + \frac{16}{8} - 11 \times 2^3$$

Now, we look at the exponents. The first term with an exponent is $3^2 = 9$, and the second one is $2^3 = 8$:

$$9 + \frac{16}{8} - 11 \times 8$$

Now we do the multiplication and division from left to right ($\frac{16}{8} = 2$) and ($11 \times 8 = 88$), to get

$$9 + 2 - 88$$

and finally, adding and subtracting from left to right, $11 - 88 = -77$.

Example 4.6, if not done in the correct order of operations, will yield several incorrect answers, and if this is a multiple-choice question, you'd better believe every one of them will be a choice.

Scientific Notation

Scientific notation uses powers of 10 to write numbers that are either too big or too small to be conveniently written as decimals. A number in scientific notation is the product of a number whose absolute value is less than 10 (in integer or decimal form) times a power of 10. Since the number part has an absolute value less than 10, it has only one digit to the left of the decimal point. The power of 10 indicates how many places the decimal point was moved from its original place in a purely decimal number. Scientific notation makes adding, subtracting, multiplying, and dividing very large and very small numbers much simpler.

There are two steps to convert to scientific notation:

1. Move the existing decimal point so that it is after the first digit. In scientific notation, the decimal number is between 1 and 10 (but not including 10).

2. Count the number of spaces the decimal point was moved. This is the power of 10. If the decimal point is moved to the left, the power of 10 is positive; if the decimal point is moved to the right, the power of 10 is negative.

Use the calculator to convert from scientific to decimal notation.

CALCULATOR

SCIENTIFIC NOTATION

The important keys for scientific notation are the $\boxed{x10^n}$ key, scrolling $\boxed{\updownarrow\!\leftrightarrow}$, the toggle key $\boxed{<>}$, and $\boxed{enter}$.

For example, to change 2.75×10^6 to a whole number, enter 2.75, press the $\boxed{x10^n}$ key, 6 (the value for n), and $\boxed{enter}$. The result, 2,750,000, shows as the answer. Likewise, for 2.75×10^{-3}, enter 2.75, press the $\boxed{x10^n}$ key, $\boxed{(-)}$ 3 (the value for n), and $\boxed{enter}$. The result, 0.00275, shows as the answer. So it works for both positive and negative exponents, but you must use the $\boxed{(-)}$ key on the keyboard for negative exponents.

For some large numbers with lots of zeros, such as 52,000,000,000,000,000,000, we can lose sight of how big that number is, having to count all those zeros, much less coming up with whether it is quadrillions or even larger. If we want to know how many miles there are in a light-year,* and we know that the distance from Earth to the sun (called an Astronomical Unit, or AU) is 1 AU = 93,000,000 miles, and that 63,000 AU = 1 light-year, the calculation (93,000,000 × 63,000) can become cumbersome. The use of scientific notation allows us to do calculations in a much more streamlined way. We can write 93,000,000 as 9.3×10^7 and 63,000 as 6.3×10^4, so the number of miles in a light-year is $(9.3 \times 10^7)(6.3 \times 10^4)$, which is $(9.3 \times 6.3) \times 10^{11} = 58.59 \times 10^{11}$, or, in scientific notation, 5.859×10^{12} miles.

This was a calculation with fairly "small" large numbers. Imagine how difficult it would be to keep track of all the zeros if each number were a billion times larger. With scientific notation, we just follow the rules for exponents: add exponents in multiplication or subtract them in division.

Scientific notation works just as well with extremely small numbers, such as the weight of an alpha particle, which is 0.000,000,000,000,000,000,000,000,006,645 kilograms. How much easier is calculation with this number if it is expressed in scientific notation as 6.645×10^{-27}?

Adding (or subtracting) two numbers in scientific notation can be done easily only if they have the same base and same exponent (just like all numbers with exponents). Since all numbers in scientific notation have the same base (10), we just have to deal with the exponents. The powers of 10 have to be the same so that the place values of the decimal parts of the numbers are in the right place. This often involves taking one of the numbers out of scientific notation. For example, the addition $3.8 \times 10^5 + 6.2 \times 10^8$ cannot be done as it is, but by converting both numbers to a value times 10^5, we have $3.8 \times 10^5 + 6200 \times 10^5 = 6203.8 \times 10^5$. Then we have to convert it back to scientific notation, so the answer is 6.2038×10^8.

Example 4.7.

The population of the United States is 3.16×10^8 and the total population in North America is 5.15×10^8. The population of the United States is $\left\{ \begin{array}{c} \text{less than} \\ \text{equal to} \\ \text{more than} \end{array} \right\}$ the population of all the other countries in North America.

Answer 4.7.

More than. The population of all the other countries in North America is what is left when the U.S. population is subtracted from the total North American population. When you subtract numbers in scientific notation, you must be using the same powers of 10. Fortunately they are both $\times 10^8$ here, so the total population of the non-U.S. countries is $5.15 \times 10^8 - 3.16 \times 10^8 = 1.99 \times 10^8$. The population of the United States is more than this number.

---

*A light-year is the distance traveled by a beam of light in one year; it is a measure of distance, not time.

Example 4.8.

Let's now compare the U.S. population to the population of China. The U.S. population is 3.16×10^8 and the population of China is 1.35×10^9. How many times more people does China have than the United States?

Answer 4.8.

4.27. "How many times more people" implies that there is a factor by which the number of people in China is larger than the number of people in the United States. That ratio would be

$$\frac{1.35 \times 10^9}{3.16 \times 10^8} = 0.427 \times 10^1 = 4.27,$$

or China has 4.27 times more people than the United States.

Exercises

1. Which of the following expressions is not equivalent to 100^{24}?

   A. $(100^6)^4$

   B. $(100^4)^6$

   C. $100^4 \cdot 100^6$

   D. $(10^2)^{24}$

2. Consider the value of $7^2 - (3 \times 8) \div 4 + \sqrt{4 \times 9}$.

   a. Which of the following operations would not be used?

   A. squaring 7

   B. taking the square root of 36

   C. subtracting 24 from 49

   D. dividing 24 by 4

   b. The value of the expression is ☐.

3. Nanotechnology deals with technology at the molecular scale. The prefix *nano-* means one-billionth, and a nanometer is 10^{-9} of a meter, or 3.937×10^{-8} of an inch. To give another perspective on how small a nanometer is, a nanometer is of the order of 10^{-5} to 10^{-4} the width of a human hair or, said another way, a human hair is approximately 90,000 nanometers wide. Write this last number in scientific notation.

    A. 9.0×10^4

    B. 9.0×10^{-4}

    C. 9.0×10^5

    D. 9.0×10^{-5}

4. Which of these numbers is not in scientific notation?

    A. 2.34×10^{-34}

    B. -2.34×10^{-34}

    C. 56.7×10^{100}

    D. 8.43×10^{10000}

5. Which of the following is the same as 65,395,000,000,000,000?

    A. 65.395×10^{16}

    B. 6.5395×10^{16}

    C. 6.5395×10^{-16}

    D. 6.5395×10^{17}

6. The value of $3 \times \sqrt{6} \times \sqrt{\dfrac{2}{3}}$ is $\begin{Bmatrix} \text{more than} \\ \text{equal to} \\ \text{less than} \end{Bmatrix}$ 6.

7. Simplify the radicals and combine: $\sqrt{12} - \sqrt{18} + \sqrt{50} - \sqrt{27}$.

    A. $2\sqrt{3} - 3\sqrt{2} + 5\sqrt{2} - 3\sqrt{3}$

    B. $2\sqrt{2} - \sqrt{3}$

    C. $-2\sqrt{2} + \sqrt{3}$

    D. $\sqrt{6}$

8. In scientific notation, the exponent of the product $(5.70 \times 10^6)(2.50 \times 10^{-4})$ is $\begin{Bmatrix} 2 \\ 3 \\ -2 \end{Bmatrix}$.

9. Simplify: $\sqrt{6} \times \sqrt{24} \times \sqrt[3]{27}$.

    A. $3\sqrt{6} \times 3$

    B. $9\sqrt{6}$

    C. 36

    D. cannot be simplified

10. Simplify: $\sqrt{1 + \dfrac{16}{25}}$.

    A. $1\dfrac{4}{5}$

    B. $\dfrac{\sqrt{41}}{5}$

    C. $\sqrt{1 + \dfrac{4}{5}}$

    D. cannot be simplified

11. Which of the following is not equal to 0.0856?

    A. 85.6×10^{-3}

    B. 8.56×10^{-2}

    C. 0.856×10^{-1}

    D. 0.00856×10^{2}

12. If a is an odd integer greater than 1, the value of $(-1)^a - 1 = \left\{ \begin{array}{c} 2 \\ 0 \\ -2 \end{array} \right\}$

13. The value of $(-27)^{\frac{2}{3}}$ is $\boxed{}$.

14. What is the value of $16^{-\frac{1}{2}}$?

    A. 4

    B. -4

    C. $\dfrac{1}{4}$

    D. $-\dfrac{1}{4}$

15. The value of $\sqrt{325}$ to the nearest integer is

    A. 13

    B. 18

    C. 60

    D. 163

Solutions

Answer 1. (C) $100^4 \cdot 100^6$. When multiplying two numbers with the same base, the exponents are added, so $100^4 \cdot 100^6 = 100^{10}$. Answer choices (A) and (B) both have a power of 100 raised to a power, so the exponents are multiplied. In answer choice (D), the value of the quantity in parentheses is 100.

Answer 2. a. (C) Subtracting 24 from 49. In the order of operations, multiplication and division would take place before subtraction, so $7^2 - (3 \times 8) \div 4$ would be equivalent to $49 - 6$.

b. 49. Using order of operations, $7^2 - (3 \times 8) \div 4 + \sqrt{4 \times 9} = 7^2 - 24 \div 4 + \sqrt{4 \times 9} = 49 - 24 \div 4 + 6 = 49 - 6 + 6 = 49$.

Answer 3. (A) 9.0×10^4. The decimal point is moved four places to the left.

Answer 4. (C) 56.7×10^{100}. The number part must be less than 10. Answer choice (B) is okay since the absolute value of the number part must be between 1 and 10 (including 1 but not including 10), so negative numbers are allowed. Answer choice (D) is okay since there is no limit on the size of the exponent.

Answer 5. (B) 6.5395×10^{16}. Count how many spaces the decimal point is moved to the left to determine the power of 10. Answer choice (A), although not in scientific notation, would be correct if the exponent was 15. Answer choice (C) has a negative exponent and should be eliminated right away.

Answer 6. Equal to. The calculation is $3 \times \sqrt{6} \times \sqrt{\dfrac{2}{3}} = 3 \times \sqrt{\dfrac{12}{3}} = 3 \times \sqrt{4} = 3 \times 2 = 6$.

Answer 7. (B) $2\sqrt{2} - \sqrt{3}$. Answer choice (A) simplifies each term but doesn't combine them.
$\sqrt{12} - \sqrt{18} + \sqrt{50} - \sqrt{27} = \sqrt{4 \times 3} - \sqrt{9 \times 2} + \sqrt{25 \times 2} - \sqrt{9 \times 3} = 2\sqrt{3} - 3\sqrt{2} + 5\sqrt{2} - 3\sqrt{3} = 2\sqrt{2} - \sqrt{3}$.

Answer 8. 3. Due to the commutative property of multiplication, $(5.70 \times 10^6)(2.50 \times 10^{-4}) = (5.70 \times 2.50) \times (10^6 \times 10^{-4}) = 14.25 \times 10^2 = 1.425 \times 10^3$.

Answer 9. (C) 36. Combine the square roots and evaluate: $\sqrt{6} \times \sqrt{24} \times \sqrt[3]{27} = \sqrt{144} \times 3 = 12 \times 3 = 36$. Another way to simplify the expression is $\sqrt{6} \times \sqrt{24} \times \sqrt[3]{27} = \sqrt{6} \times 2\sqrt{6} \times 3 = 6 \times 2 \times 3 = 36$.

Answer 10. (B) $\dfrac{\sqrt{41}}{5}$. You can separate products but not sums under the radical sign, so $\sqrt{1 + \dfrac{16}{25}} \neq \sqrt{1} + \sqrt{\dfrac{16}{25}}$. Answer choices (A) and (C) are wrong. But since $1 + \dfrac{16}{25} = \dfrac{25}{25} + \dfrac{16}{25} = \dfrac{41}{25}, \sqrt{1 + \dfrac{16}{25}} = \sqrt{\dfrac{41}{25}} = \dfrac{\sqrt{41}}{\sqrt{25}} = \dfrac{\sqrt{41}}{5}$.

Answer 11. (D) 0.00856×10^2. Move the decimal points as indicated in each of the answer choices.

Answer 12. -2. The value of -1 raised to an odd power always equals -1 because an odd number of minuses multiplied together is always minus. Thus, $(-1)^a - 1 = -1 - 1 = -2$.

Answer 13. 9. The fractional exponent $\dfrac{2}{3}$ means "square the cube root," so $(-27)^{\frac{2}{3}} = \left(\sqrt[3]{-27}\right)^2 = (-3)^2 = 9$.

Answer 14. (C) $\dfrac{1}{4}$. A negative square root means reciprocal, and the fractional exponent $\dfrac{1}{2}$ means "square root," so $16^{-\frac{1}{2}} = \dfrac{1}{\sqrt{16}} = \dfrac{1}{4}$.

Answer 15. (B) 18. Use the calculator to find that $\sqrt{325} = 18.028$, which is 18 to the nearest integer. This problem can also be done by squaring each answer or after eliminating answer choices (C) and (D) because they are clearly too large.

Chapter 5

Algebra— Not a 4-Letter Word

The very mention of the word *algebra* strikes fear in the hearts of many students. Students have said, "Who cares what *x* is?" and "What does this have to do with the real world?" Well, algebra is something you use every day—you just don't recognize that you are doing algebra. If I asked you, "How much change would you get from a $10 bill for a purchase of $7.20?" you would probably be able to come up with the answer ($2.80) pretty quickly. But if I asked you to solve the equation $x + 7.20 = 10$ for *x*, maybe the answer wouldn't come as fast. It's the same thing. In fact, a number of the problems you already have done in this book have been algebra problems, but they were asked as "what is?" type questions, so you probably didn't notice that they were the same as "$x =$" type equations.

By learning a few basics, as well as how to interpret some key words, algebra should become less daunting. More than half of the GED® math test involves algebra, so it's pretty important to be able to understand the problems. Once you understand what is being asked, the arithmetic to get the correct answer is fairly easy. This chapter introduces basic concepts and vocabulary and then presents lots of examples to help you feel more confident in solving algebraic problems.

Setting up Equations

Algebra involves **equations**. The word *equations* has the same root as *equal*, and that is what equations are all about. Whatever is on one side of the equals sign must equal what is on the other side. So if you add something to one side, you must also add it to the other side. The same is true for subtracting, multiplying, dividing, taking roots, or raising to powers.

Without using x's, we can see this is true by considering the equation $8 + 2 = 6 + 4$. If we decide to add 3 to the left side, we must add 3 to the right side of this equation or it won't be true: $8 + 2 + 3 = 6 + 4 + 3$. It's that simple. If instead we subtracted 3 from each side, $8 + 2 - 3 = 6 + 4 - 3$, also true.

The usual letters used for the unknown quantities, which are called **variables**, are x and y, although any letters will work as well. Remember, they are just taking the place of a real value that we want to find. Treat them like you would treat a number. Thus, for two numbers x and y, their sum is $x + y$; their difference is $x - y$; their product is $x \times y$ (often written without the $\times$ sign as just xy); and their quotient is $x \div y$ (also written as $\frac{x}{y}$).

The only other thing we need to know is that solving an algebraic equation means getting the unknown quantity on one side of the equation by itself so we can get a value for it. To do that, we must get rid of anything that is on the same side of the equation as the unknown. We do that by using the "opposite" operation, called the **inverse**, and remembering to do that operation on both sides of the equals sign. The inverse of addition is subtraction (and vice versa), and the inverse of multiplication is division (and vice versa).

Let's consider the equation $\qquad x + 5 = 7$.

We must get rid of the 5 on the same side of the equation as x. The opposite of the $+5$ on the left side is -5, so we must subtract 5 from both sides of the equation. We thus get

$$x + 5 - 5 = 7 - 5.$$

The $+5$ and -5 cancel each other out on the left side, leaving x alone. So we now have

$$x = 7 - 5,$$

and the answer is $\qquad x = 2$.

As a check, replace x by 2 in the original equation: $2 + 5 = 7$. It checks.

Similarly, if a quantity is subtracted from the unknown on one side of the equation, add that quantity to both sides to get a value for the unknown. The addition will cancel the subtraction:

$$x - 8 = 4$$
$$x - 8 + 8 = 4 + 8$$
$$x = 12.$$

Again, check this answer in the original equation: $12 - 8 = 4$.

For multiplication, the inverse is division, and for division, the inverse is multiplication. We just have to remember to do the same thing to both sides of the equation. For example, find y if

$$3y = 12.$$

We divide both sides of the equation by 3: $\dfrac{3y}{3} = \dfrac{12}{3}$,

$$y = 4.$$

The check for this is $3(4) = 12$.

Similarly, for $$\dfrac{x}{4} = 5,$$

we multiply both sides by 4: $$\dfrac{x}{4} \times 4 = 5 \times 4$$

$$x = 20.$$

Check: $\dfrac{20}{4} = 5$.

Of course, the GED® test wants you to be able to set up these kinds of equations. The solutions to equations that have only one variable that isn't raised to a power won't be much more difficult than the samples given above; the catch is that you have to find out what the equation is—it won't always be given to you. You should be able to represent a situation by algebra so you can find whatever value is asked for.

Luckily, if you know the "translations" of some words, this task will be easier.

1. The words *is* (or any variation of it), *cost*, and *is the same as* mean *equals* and should be substituted by an equals sign.

2. The words *what* or *how much* (or a similar question) mean the unknown, so replace them with the variable—x, y, or whatever letter you choose. The equation you construct will find the value for this variable.

3. *Sum, plus, in all,* or *combined* mean addition.

4. *Difference, less than, how much more, exceeds,* or *minus* mean subtraction.

5. *Product, times, area,* or *of* (such as *half of*) mean multiplication.

6. *Quotient*, *distribute*, or *per* mean division.

7. *Decreased by* often indicates subtraction.

8. *Increased by* often indicates addition.

Let's look at a few examples of situations that involve algebra. Notice that the unknown quantity is not always *x*.

Example 5.1.

What is the difference between 32 and 28?

Answer 5.1.

Translation: (What) (is) (the difference between) 32 and 28?

x = subtract 32 and 28

$x = 32 - 28$

$x = 4.$

Example 5.2.

What number is 7 less than 12?

Answer 5.2.

Translation: (What number) (is) 7 less than 12?

n = 12 subtract 7

$n = 12 - 7$

$n = 5.$

Example 5.3.

What number is 7 more than 12?

Answer 5.3.

Translation: (What number)   (is)   7   (more than)   12?

                        p =   7   +   12

$$p = 7 + 12$$
$$p = 19.$$

Example 5.4.

What number decreased by 7 is 12? (Notice the difference between Example 5.3 and this example.)

Answer 5.4.

Translation: (What number)   (decreased by)   7   (is)   12?

                        x $-$ 7   =   12

$$x - 7 = 12$$

Add 7 to both sides to get $x = 19.$

Example 5.5.

Seven times what number is 21?

Answer 5.5.

Translation: (Seven)   (times)   (what number)   (is)   21?

                  7            $\times$ x =   21

$$7x = 21$$

Divide both sides by 7: $x = 3.$

Note that in algebra, the multiplication sign is often omitted, so $7x$ is the same as $7 \times x$.

Most problems on the GED® test will not be as straightforward as Examples 5.1−5.5. You may have to think through what is being asked. The arithmetic, though, will not be difficult.

Example 5.6.

John and Jill want to buy a house. They have $30,000 in their savings account. If they use $12,000 to pay off Jill's student loan, how much of a down payment can they afford?

Answer 5.6.

Perhaps you already can see that the answer is $18,000. If you can, you really have used algebra to get it. The equation says in symbols that the amount for the down payment is the difference between their savings and their loan obligation, or

$$x = \$30,000 - \$12,000.$$

If we do the subtraction, we can find the value of $x = \$18,000$ for a down payment.

Now let's look at a similar problem.

Example 5.7.

John says to Jill, "If we take $18,000 for the down payment from our savings, we would have only $12,000 to pay toward your student loan." How much do they have in savings?

Answer 5.7.

In symbols, the equation is something like

$$y - \$18,000 = \$12,000.$$

We still want to find the unknown; this time we called it y, so we want y to be alone on one side of the equation. Then if we evaluate everything on the other side, we will have a value for y. If we add $18,000 to each side, we get

$$y - \$18,000 + \$18,000 = \$12,000 + \$18,000$$

$$y = \$30,000.$$

Example 5.8.

Jimmy has seven marbles. Tommy has eight times as much as Jimmy. How many marbles does Tommy have?

Answer 5.8.

Translation: (Tommy) (has) (eight) (times as much as) (what Jimmy has)

$$T \quad = \quad 8 \quad \times \quad 7 \text{ (from the first sentence)}$$

$$T = 8 \times 7 = 56 \text{ marbles.}$$

Example 5.9.

Jane divided her collection of 24 dolls evenly among her three younger cousins. How many dolls did each cousin get?

Answer 5.9.

The 24 dolls were divided among three cousins, so the equation is

$$x = 24 \div 3 = 8 \text{ dolls.}$$

Working with Algebraic Expressions

Equations with one variable don't always involve just one operation, as those in Examples 5.1 to 5.9 do. Consider the following equation:

$$3x + 7 = 13.$$

There are two operations on the side of the equation that contains the variable: multiplication and addition. Each must be eliminated to leave the variable alone on the left side. First, every term without a variable on the left side must be taken away. Since the 7 is added, we use the inverse operation and subtract 7 from *both* sides. This leaves us with

$$3x = 6.$$

To now get rid of the 3 that is multiplying the x, we must divide both sides of the equation by 3 because division is the inverse of multiplication, and the solution is

$$x = 2.$$

So we used the same rules as discussed earlier to get the variable alone on one side of the equation, but here we had to use two rules for the two operations. In the next section, we discuss algebraic expressions and combining like terms, which present a slightly more complex situation in equations with more than one variable, but the above rules remain the same.

Adding and Subtracting Algebraic Expressions

An **algebraic expression** is any combination of numbers and variables. It can, in fact, involve many variables. Examples of algebraic expressions are $3x$, $3x + 4y$, $5xy$, and $6a + 5b - 4c$. The numbers in front of the variables are called **coefficients**, and each part of the expression (separated by $+$ or $-$ signs) is called a **term**.

The operations of addition and subtraction that we do to combine numbers are also done with algebraic terms, but algebraic terms can be combined only if they are **like terms**. Like terms have exactly the same variables to the same powers.

For example, $3x$ and $5x$ are like terms because they both have x to the first power (remember, we don't have to write the exponent 1). We can add them ($3x + 5x = 8x$) and subtract them ($3x - 5x = -2x$). In contrast, we cannot add or subtract $3x$ and $5xy$ or $3x$ and $3x^2$ because their variables are not identical.

Example 5.10.

Write in simplest form: $4x + 5xy - 6y + 6xy - 2x - xy$.

Answer 5.10.

We can combine only the terms $4x$ and $-2x$, and the terms $5xy$, $6xy$, and $-xy$, but the $-6y$ remains uncombined because there isn't another term with only y. The expression thus becomes

$$4x + 5xy - 6y + 6xy - 2x - xy = 4x - 2x + 5xy + 6xy - xy - 6y = 2x + 10xy - 6y.$$

If we think of each of the variables to be a different fruit, $x =$ apples, $y =$ oranges, and $xy =$ bananas, this makes perfect sense. That is what is meant by the expression, "You can't compare apples and oranges"—they are not the same. Likewise, x and any of its powers (x^2, x^3, x^{10}) are not like terms, either, even though they all involve x. Their powers must also match.

Multiplying and Dividing Algebraic Expressions

Algebraic expressions do not have to have like terms for multiplication or division. For multiplication, simply multiply the coefficients and then multiply the variables. Therefore, $(2x)(3y) = 6xy$. Remember that parentheses are one way to indicate multiplication, and it is especially used in algebra. Likewise, $(2x)(3x)(4y) = 24x^2y$, remembering that $(x)(x) = x^2$.

For division, simply divide the coefficients and then divide the variables, remembering that in division, like factors in the numerator and denominator cancel and a negative exponent means reciprocal. Thus, $15xy \div (-3x) = -5y$ because the x's cancel.

This brings up the topic of the next section, which presents rules for exponents when multiplying and dividing expressions.

Exponent Rules

As we saw in Chapter 4 with numbers, there are three important rules about exponents. For these rules, the exponents can be different, but the bases (the variable being raised to a power) must be the same. That means that when multiplying or dividing, you can combine x^2 and x^4 (same base, x) but not x^2 and y^2, and not x^2 and y^4 because x and y are different bases.

1. When multiplying one power of a variable by another power of the same variable, add the exponents. Thus, $x^2(x^3) = x^5$. When we discussed powers in Chapter 4, we said the exponent tells how many times the factor appears, so $x^2 = x \times x$ (2 x's), and $x^3 = x \times x \times x$ (3 x's). When we multiply $x^2(x^3)$, we get $(x \times x) \times (x \times x \times x)$, or 5 x's, which we write as x^5, adding the exponents.

2. When dividing one power of a variable by another power of the same variable, subtract the exponents (top minus bottom). Thus, $\dfrac{x^5}{x^2} = x^3$. Remember that when we divide, we can cancel factors in the numerator and denominator of the fraction, so this is

$$\frac{x \times x \times x \times \cancel{x} \times \cancel{x}}{\cancel{x} \times \cancel{x}} = x \times x \times x = x^3.$$

3. When raising a power to a power, multiply the exponents. Thus, $(x^3)^2 = x^6$. Using the same reasoning as above, we have $(x \times x \times x) \times (x \times x \times x) = x^6$ (6 x's).

When dividing expressions with exponents, which involves subtracting the exponents, it is possible to get a **negative exponent**. Remember that a negative exponent indicates a **reciprocal**. Thus, $\dfrac{x^3 y^4}{x^5 y} = x^{-2} y^3 = \dfrac{y^3}{x^2}$.

A quicker way to do the same problem is to subtract the exponents without regard to sign and to put the answer (with a positive exponent) in the part of the fraction that has the larger exponent. Thus, for the same problem, the thinking is as follows. The exponents of x are 3 and 5, the difference is 2 and the larger exponent is in the bottom of the fraction, so that's where x^2 goes. For the y's, $4 - 1$ is 3, so y^3 goes in the top. Right away you can write the answer, $\dfrac{x^3 y^4}{x^5 y} = \dfrac{y^3}{x^2}$. This works really well when there are a lot of variables. And it took lots longer to explain the solution here than to actually work the solution.

Example 5.11.

Reduce to lowest terms: $\dfrac{(7xy)(6xy)}{21x^3y^4}$.

Answer 5.11.

$$\frac{2}{xy^2} \cdot \frac{(7xy)(6xy)}{21x^3y^4} = \frac{42x^2y^2}{21x^3y^4} = \frac{2}{xy^2}.$$

Ratio and Proportion

As shown in Chapter 3, which discussed fractions, a **proportion** is the way to express that two ratios are equal. Ratios are often written as fractions, $\dfrac{x}{y}$, stated as "the ratio of x to y." The problem given in Chapter 3 was "If the ratio of men to women in the workplace is 3 to 7, how many men are in an office with 70 women?" Algebraically, this is set up as the equalities of the two ratios (which is the definition of proportion), with a variable taking the place of the unknown quantity, or $\dfrac{3}{7} = \dfrac{x}{70}$ for this problem. You probably see that $x = 30$ here, but the answer isn't always so obvious. However, the solution method is not complicated using simple algebra.

In a proportion, set up the ratios as above with equal fractions and a variable used for the quantity to be determined. Then use **cross-multiplication**, which, as the name implies, means multiplying in the form of a cross (**X**), making the two products equal. For the problem above, $\dfrac{3}{7} \diagup\!\!\!\!\diagdown \dfrac{x}{70}$, this means $3 \times 70 = 7x$, or $7x = 3 \times 70$. Either way is correct—just remember which numbers are the products. This is then solved algebraically as $7x = 210$, or $x = 30$ men.

Example 5.12.

Evonne can type 90 words per minute on her computer keyboard. How many words can she type in an hour (assuming no breaks or fatigue)?

Answer 5.12.

The first ratio is $\dfrac{90 \text{ words}}{1 \text{ minute}}$. An hour is 60 minutes, and we want to find how many words in an hour, so the second ratio is $\dfrac{x \text{ words}}{60 \text{ minutes}}$. It is very important that the units on both sides of a proportion are the same. These two ratios are equal, representing the rate at which Evonne can type, so the equation is $\dfrac{90 \text{ words}}{1 \text{ minute}} = \dfrac{x \text{ words}}{60 \text{ minutes}}$. Cross-multiplication then gives us $x = (90)(60) = 5400$ words. You might have gotten that same answer without all of the math, but in reality you were using algebra in your head.

As expected, the numbers won't always be so straightforward on the GED® test, but as long as you remember to set proportions up as fractions, to use cross-multiplication, and to use some basic algebra to set up the equation to be solved, your calculator will do the "dirty" work for you when the numbers get large.

Example 5.13.

Suppose Mike walks a half mile in 12 minutes. How many miles can he walk in 3 hours?

Answer 5.13.

The first ratio is $\dfrac{.5 \text{ mile}}{12 \text{ minutes}}$. The second ratio (note that we have to have the same units, so we use 3 hours is $3 \times 60 = 180$ minutes) is $\dfrac{x \text{ miles}}{180 \text{ minutes}}$. They both represent Mike's rate of walking, so they are equal: $\dfrac{.5 \text{ mile}}{12 \text{ minutes}} = \dfrac{x \text{ miles}}{180 \text{ minutes}}$. Cross-multiplication then gives us

$$12x = (.5)(180)$$

$$x = \frac{(.5)(180)}{12}$$

$$x = \frac{90}{12} = 7.5 \text{ miles.}$$

Evaluating Algebraic Expressions

To come up with a value for an algebraic expression, you must be given values for each of the variables. A typical problem of this type is "Evaluate the expression $3x - 4y$ if $x = 2$ and $y = 5$." So you rewrite the expression, putting a 2 in for the x and a 5 in for the y; evaluate by using the order of operations (PEMDAS, or Parentheses, Exponents, Multiplication, Division, Addition, and Subtraction), as discussed in Chapter 4, and you are done. In this case, the answer is $3x - 4y = 3(2) - 4(5) = 6 - 20 = -14$.

Not all evaluation problems are that straightforward, which is why you must use the order of operations in the final evaluation. The expression $x^2 - xy \div 20 \times 7$ will give some very interesting (but wrong) answers if the order of operations isn't followed. This expression, after substituting the same values of $x = 2$ and $y = 5$ is

$$2^2 - 2(5) \div 20 \times 7 = 2^2 - 10 \div 20 \times 7 = 4 - 10 \div 20 \times 7 = 4 - .5 \times 7 = 4 - 3.5 = 0.5.$$

Notice that the multiplication and division must be done in left-to-right order. If they aren't, you may end up dividing 10 by 140 and get a completely different answer.

Solving Linear Equations

So far, we have talked about solving equations with one unknown to the first power, so the answer is one number. Equations with two unknowns, usually x and y each to the first power, are known as linear equations because the x and y values that make these types of equations true form a line when graphed on an xy-coordinate system.

First, let's look at this coordinate system, called the **Cartesian coordinate system**, shown on the next page. The **x-axis** runs horizontally and the **y-axis** runs vertically. The axes (plural of *axis*) divide the coordinate system into four **quadrants**, usually labeled counterclockwise with Roman numerals. The point where the axes meet is called the **origin**.

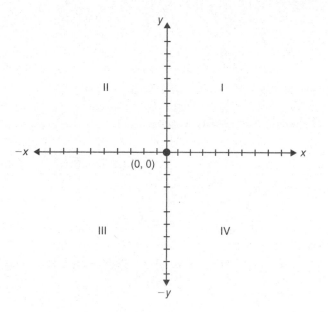

Any point on the Cartesian coordinate plane has an x value and a y value and is labeled by these values as the **ordered pair** (x, y). This is called an ordered pair because the order matters. The x value always comes before the y value. The origin has the coordinates $(0, 0)$.

To determine the coordinates for any point, follow the point up or down to the x-axis to read the x value, and across to the y-axis to read the y value. To plot a point, reverse this process and find where the x value and the y value meet.

The distance between two points on a Cartesian coordinate plane is found by the formula for the Pythagorean Theorem, which is discussed in Chapter 7.

If you find two points on the graph, you can draw a straight line through them, and whatever the equation is for that line, it is true for every point on that line. It is always a good idea to plot a third point before you draw the line, though, because if one of the original two points is in error, that will show up when the three points don't line up. The third point acts as a check.

So how do we get these points? Let's use the equation $y = x - 2$ as an example. Remember that we want three points. We can choose any value for either x or y, then substitute it in the original equation and solve the resulting equation to get a value for the other variable. Usually, this is done in a table and usually the first two values we try are $x = 0$ and $y = 0$ because they simplify the equation. Then we substitute any value of x or y for the third point and solve the resulting equation to find the other variable, as shown in Example 5.14.

Example 5.14.

Plot the line for $y = x - 2$.

Answer 5.14.

First we set up a table of three points to plot.

x	y	Explanation
0	−2	When x = 0, y = −2. Plot (0, −2) on the graph.
2	0	When y = 0, x = 2. Plot (2, 0) on the graph.
1	−1	To show that any point that satisfies the equation will work, let's use x = 1 instead. Then y = −1 satisfies the equation, and the point is (1, −1). It should fall on the line with the other two points.

The plot of this line is shown on the graph below. Notice that all three points that we determined are on this line.

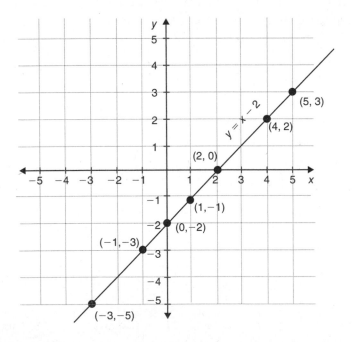

Any straight line goes on forever in either direction, even though only a part of it is shown on the graph. Let's check out some of the other points on the line, now that we can read them from the graph. There are plenty (actually an infinite number) of points on the line, but let's just check a few that are whole numbers, such as $(-3, -5)$, $(-1, -3)$, $(4, 2)$, and $(5, 3)$. Remember that the points are written in the following order: (x, y). We substitute these values into the original linear equation $y = x - 2$, and we see that they all satisfy the equation:

$$(-3, -5): -5 = (-3) - 2$$

$$(-1, -3): (-3) = (-1) - 2$$

$$(4, 2): 2 = 4 - 2$$

$$(5, 3): 3 = 5 - 2$$

So our points for this line are $(-3, -5)$, $(-1, -3)$, $(0, -2)$, $(1, -1)$, $(2, 0)$, $(4, 2)$, and $(5, 3)$. Let's put these in a table:

x	-3	-1	0	1	2	4	5
y	-5	-3	-2	-1	0	2	3

Can you see a pattern? Without doing the algebra, can you see what the value of y would be when $x = -2$? How about the value of x when $y = 1$? Check the line to see whether your answers are right. The pattern appears to be that for every change of $+1$ in the x value, there is a corresponding change of $+1$ in the y value.

Slope

This "pattern" tells us the **slope** of the line between two points (x_1, y_1) and (x_2, y_2), which is defined as $m = \dfrac{\text{the change in } y}{\text{the change in } x}$, or $m = \dfrac{y_2 - y_1}{x_2 - x_1}$. This formula is on the formula sheet on the GED® test. If we multiply both sides of this equation by $(x_2 - x_1)$, we get the **point-slope form**, or $(y_2 - y_1) = m(x_2 - x_1)$, which is also on the formula sheet of the GED® test. So if the only information you have for a line is a point on the line and the slope of the line, you can find the equation of the line.

For example, for the equation of a line that passes through the point $(4, 2)$ with a slope of 3, let the other point be (x, y). Then the change in y can be written as $(y - 2)$ and the change in x can be written as $(x - 4)$. Substituting the values into the point-slope form, $(y_2 - y_1) = m(x_2 - x_1)$, we get

$$(y - 2) = 3(x - 4)$$

$$y - 2 = 3x - 12$$

$$y = 3x - 12 + 2$$

$$y = 3x - 10.$$

An easier way to remember slope is "rise over run," or $\dfrac{\text{rise}}{\text{run}}$; between any two points, the rise (change in y) divided by the run (change in x) is the same, and this value is the slope. If the slope is positive, it goes from the lower left to the upper right on the graph, as the line on the graph for Example 5.14 does. The slope is $\frac{1}{1} = 1$.

For a negative slope, let's look at the equation $2x + y = 6$. Remember that we want three points.

Example 5.15.

Graph the line $2x + y = 6$.

Answer 5.15.

x	y	Explanation
0	6	When $x = 0$, $y = 6$. Plot (0, 6) on the graph.
3	0	When $y = 0$, $2x = 6$, so $x = 3$. Plot (3, 0) on the graph.
2	2	When $x = 2$, the equation is $4 + y = 6$, so $y = 2$. Plot (2, 2) on the graph. It should fall into line with the other two points.

The graph is shown below.

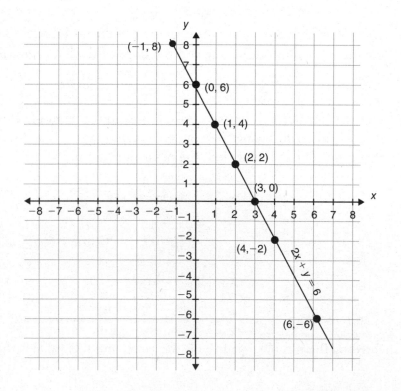

We can read the following points on this line: $(-1, 8)$, $(0, 6)$, $(1, 4)$, $(2, 2)$, $(3, 0)$, $(4, -2)$, and $(6, -6)$. Let's put this in a table:

x	−1	0	1	2	3	4	6
y	8	6	4	2	0	−2	−6

Again, can you see a pattern? Without doing the algebra, can you see what the value of y would be when $x = 5$? How about the value of x when $y = -8$? Check the line to see whether your answers are right. The pattern seems to be that, for every change of $+1$ in the x value, there is a corresponding change of -2 in the y value, so the slope is $\frac{-2}{1} = -2$. For a negative slope, the line goes from the upper left to the lower right.

But can we determine the slope of a line without actually plotting it? Yes we can, by putting the equation in what is called **slope-intercept form**, $y = mx + b$. This equation is on the formula sheet on the GED® test, but you have to know what m and b are. Just as before, m is the slope, and b is the y-intercept, which means the point where the line intercepts (crosses) the y-axis.

Let's check out the slope-intercept form for both Examples 5.14 and 5.15. For Example 5.14, $y = x - 2$ already is in slope-intercept form, with $m = 1$ (the coefficient of x), and $b = -2$, the y-intercept. The graph of this equation indeed does show a y-intercept of -2, the point $(0, -2)$.

For Example 5.15, we have to do a little algebra to put $2x + y = 6$ into slope-intercept form. We want y alone on one side of the equals sign, so we subtract $2x$ from each side.

$$2x + y = 6$$

$$y = -2x + 6,$$

from which we can read that the slope is -2 (for every increase of 2 in y, there is a corresponding *decrease* of 1 in x) and the y-intercept is $+6$. Looking at the graph, we see that this is also true.

Distance–Time Graphs

Since slope gives rate of change, we can graph the rate at which something moves by plotting time on the x-axis and distance on the y-axis. In these cases, we will assume a constant speed (rate). Suppose Terrell drives 100 miles on the highway in 2 hours. The graph would look like the following:

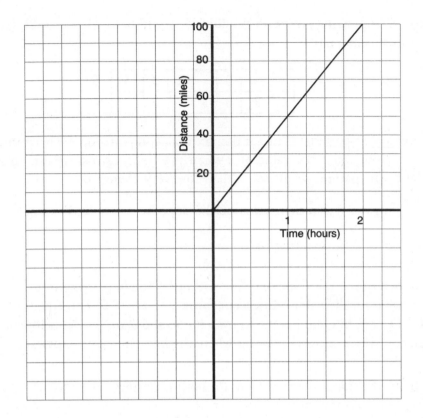

The slope of the line gives the speed at which Terrell is traveling. Be careful here because the scales are different on the two axes. On the horizontal (time) axis, each square is one-fourth of an hour, or 15 minutes, whereas on the vertical (distance) axis, each square is ten miles.

 HINT

The scales on the axes of a graph don't have to be the same, especially if they each measure a different unit. When graphing, use the scale that makes sense. When reading a graph, be sure to check the scales first. For example, if the measurement was the movement of an iceberg over hundreds of years, it wouldn't make sense to make each square on the time scale equal to one year.

The slope is still given by $m = \dfrac{y_2 - y_1}{x_2 - x_1}$, where the x's are time and the y's are distance. The slope for this graph is $m = \dfrac{100 - 0}{2 - 0} = 50$ miles per hour.

This agrees with the general equations that relate distance (d), rate (r), and time (t), which are

$$d = rt, \quad r = \frac{d}{t}, \quad t = \frac{d}{r}.$$

All three equations are saying the same thing, and if you know one, such as $d = rt$, you can get the others.

HINT

An easy way to remember $d = rt$ is by the word *dirt*. Also, if you remember that speed (rate of travel) is measured in miles/per hour, you can remember $r = \dfrac{d}{t}$.

So Terrell is traveling at a speed of $r = \dfrac{d}{t} = \dfrac{100 \text{ miles}}{2 \text{ hours}} = 50$ miles per hour, the same speed as we got from the graph.

This same relationship between slope on a graph and rate in the equation works for all constant rate problems, whether measuring words per minute on a keyboard or lightning strikes per minute in the weather report.

Unit Rates

Slope represents a specific change in one variable due to a specific change in the other variable. In other words, slope in a proportional relationship (of the form $y = mx$) represents the proportion between the variables or the rate of change (increase or decrease) between them. Specifically, a **unit rate** describes how many units of the first variable correspond to one unit of the other variable. Whatever proportion exists between the variables should be changed to a single-unit rate. In other words, if the slope is represented by $m = \dfrac{3}{2}$, that value should be changed to a multiple-unit rate with 1 as the denominator. The proportion would be

$$m = \frac{3}{2} = \frac{y}{1}.$$

Here, that would be a rate of $y = 1.5$ (for the first unit) to $x = 1$ (for the second unit). The "units" referred to here are usually different, such as miles per hour or price per ticket.

Example 5.16.

Ever since a certain tree was 10 feet tall, it has been growing at a rate of 5 feet every 3 years. A plot of the growth, with x = years, and y = tree height, shows a line with a slope of $m = \frac{5}{3}$ and a y-intercept of 10 feet. At what annual rate has the tree been growing since it was 10 feet tall?

Answer 5.16.

The rate of growth is $m = \frac{5}{3}$. To change that to a unit growth, use the proportion $\frac{5}{3} = \frac{y}{1}$, so $y = 1.67$. This means the tree is growing at a rate of 1.67 feet per year.

Parallel and Perpendicular Lines

We know from the point-slope formula that the slope of the line $y = 2x - 6$ is 2. If a line is drawn on the same grid parallel to $y = 2x - 6$, what do you think its slope will be? Since **parallel** means two lines that are an equal distance apart and go on forever and never meet or cross, it makes sense that the slopes of parallel lines are the same. In fact, $y = 2x - 6$ is parallel to all of the lines $y = 2x + 8$, $2y - 4x = 8$, and $3y = 6x$ because the slopes for all of these lines is 2. When we plot them on the same grid, as shown below, we get four parallel lines.

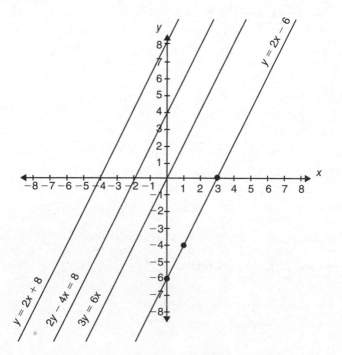

What about the line $2y = 4x - 12$? If you put this line into slope-intercept form, indeed, its slope is 2, but its y-intercept is -6, the same as our original line $y = 2x - 6$. In fact, $2y = 4x - 12$ is the same line as $y = 2x - 6$. You can see this if you divide all the terms of $2y = 4x - 12$ by 2. Also, if we try to plot $2y = 4x - 12$ on the same grid, we see that it is the same line as $y = 2x - 6$.

Perpendicular lines, in contrast, are lines that meet, or intersect, at a right angle (90°). The slopes of perpendicular lines are negative reciprocals of each other. So the line $2y = -x + 10$, which is $y = -\dfrac{1}{2}x + 5$ in slope-intercept form is perpendicular to our original line, $y = 2x - 6$ because its slope, $-\dfrac{1}{2}$, is the negative reciprocal of 2. Try plotting $2y = -x + 10$ on the same grid as we used for the parallel lines on the previous page. It is perpendicular to all of the parallel lines.

Example 5.17.

What is the slope of a line parallel to $3y = 8x - 6$?

Answer 5.17.

For a line to be parallel to another line, they must have the same slopes. The slope of the given line is found by putting it in slope-intercept form, or $y = \dfrac{8}{3}x - 2$. The slope is thus $\dfrac{8}{3}$, and that is the slope of any line parallel to it.

Example 5.18.

What is the slope of a line perpendicular to $3y = 8x - 6$?

Answer 5.18.

For a line to be perpendicular to another line, their slopes must be negative reciprocals. The slope of the given line was found in Example 5.17 to be $\dfrac{8}{3}$, and thus the slope of any line perpendicular to it must be $-\dfrac{3}{8}$. An example would be the line $8y = -3x + 16$, which is $y = -\dfrac{3}{8}x + 2$ in slope-intercept form.

Simultaneous Equations

Sometimes it is necessary to find the one point at which two linear equations are true. This requires us to find two quantities, which means we need two equations (we always need as many equations as unknowns); these equations are referred to as **simultaneous equations** or **systems of equations**.

For example, let's say that 1,000 tickets were sold to an event. Adult tickets cost $10 and children's tickets cost $4, and a total of $7,600 was collected. How many tickets of each kind were sold? Neither sentence describing the situation can give us the answer alone, but taking information from both sentences *simultaneously* does. We are looking for the number of adult tickets (let's call that a) and the number of children's tickets (let's call that c).

Then the first sentence tells us how many tickets were sold, so the total number of adult and children's tickets is

$$a + c = 1,000.$$

But that's not enough to figure out how many of each type of ticket were sold (it could be that 1 adult ticket and 999 children's tickets were sold, or 500 of each, or whatever numbers add to 1,000).

The rest of the problem gives more information so we can find out exactly how many of each type of ticket were sold. The adult tickets cost $10, so $10a$ tells how much money came in from adult ticket sales. Similarly, the children's tickets cost $4, so $4c$ tells how much money came in from children's ticket sales. If we total these amounts, we know that they must equal the $7,600 that was collected. So the second equation is

$$10a + 4c = 7,600.$$

Now we put these two equations together to find out what values of a and c will make them both true. We have several ways to do this.

Algebraically, we can find an expression for a or c from the first equation and then substitute that into the second equation and solve it. From the first equation, $a + c = 1,000$, we can subtract a from each side to get an expression for c, which is

$$c = 1,000 - a.$$

Let's substitute that for c in the second equation, and then we can figure out what a is. So the second equation becomes

$$10a + 4(1,000 - a) = 7,600$$

$$10a + 4,000 - 4a = 7,600$$

Combining like terms and subtracting 4,000 from both sides give us

$$6a = 3,600, \text{ so } a = 600 \text{ adult tickets.}$$

Then the number of children's tickets must be $c = 1,000 - 600 = 400$.

Indeed, if we substitute $a = 600$ and $c = 400$ into each of the original equations, they both are true. In fact, this situation is not true for any other combinations of ticket sales.

Solving Simultaneous Equations

This seems like a lot of work, but the hardest part was figuring out what the two equations should be. The algebra was pretty straightforward.

Probably the GED® test will give two equations and multiple choices for the answer, or a situation from which to choose the two equations from multiple choices (and not have to solve the problem at all). Let's look at those types of problems.

Example 5.19.

Which of the following values are solutions to the following system of equations?

$$x + y = 3$$

$$5x + 8y = 21$$

A. $x = -1, y = -2$

B. $x = -1, y = 2$

C. $x = 1, y = -2$

D. $x = 1, y = 2$

Answer 5.19.

(D) $x = 1$, $y = 2$. If this is a system of equations, the values for x and y must be true for both equations. So we plug in those values in both equations to see which pair of values works. Right away, just looking at the first equation, we can see that the only possible pair that is true for the first equation is (D), but we must also check that it is a solution to the second equation, just to be sure that we didn't make a mistake.

Example 5.20.

Choose the simultaneous equations you would use to solve the following problem: "Robert has 20 coins, consisting of dimes and nickels. The total value of the coins is $1.60. How many of each coin does Robert have?"

A. $n - d = 20$

$5n + 10d = 160$

B. $n + d = 20$

$5n + 10d = 160$

C. 12 dimes

8 nickels

D. $n + d = 20$

$10d - 5n = 160$

Answer 5.20.

(B) $n + d = 20$, $5n + 10d = 160$. First eliminate choice (C) because, although it is the correct answer to the problem, it is not what the question asked for, which was which pair of simultaneous equations to use to solve the problem. Check the first equation of each choice to see whether it matches the problem. Eliminate choice (A) because it doesn't match the first sentence of the problem. That leaves answer choices (B) and (D) to match "the total value of the coins is $1.60." "Total" means "add," so the correct answer is (B). Note that all of the values are changed to cents, which is a good idea in coin problems because it simplifies the math.

Example 5.21.

Which of the following equations, together with the equation $3x - 4y = 6$, is sufficient for finding the values of x and y?

    A. $3x - 6 = 4y$

    B. $x - \dfrac{4}{3}y = 2$

    C. $6x - 8y = 12$

    D. $4x - 3y = 8$

Answer 5.21.

(D) $4x - 3y = 8$. We want an equation that will intersect the first equation at just one point, which would be the solution to both equations. The best way to check this is to put each equation into slope-intercept form. The first three answer choices have the same slope and intercept as the given equation. So they are each the original equation written a different way. Answer choice (A) just rearranges the terms, (B) divides each term by 3, and (C) multiplies the equation by 2. So all three of these choices are actually the same line as the original, having infinite points in common. Only (D) will intersect with the original equation at only one point, the values of x and y.

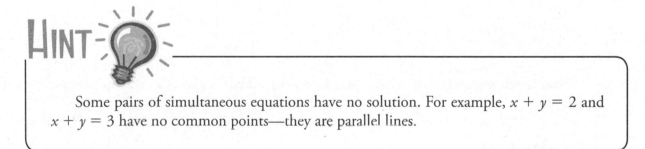

Some pairs of simultaneous equations have no solution. For example, $x + y = 2$ and $x + y = 3$ have no common points—they are parallel lines.

Graphing Simultaneous Equations

Another method for solving simultaneous equations involves graphing each equation on the same grid. Since there is only one answer and it must be true for both equations, these graphs should cross at the values that are solutions to both equations. Let's graph the equations from Example 5.19. The two equations are $x + y = 3$ and $5x + 8y = 21$.

We just have to pick three points to graph each line. First substitute $x = 0$ into each equation to get the corresponding y value. Then substitute $y = 0$ to get the x value at that point. For the third, substitute a convenient (small) value for x or y. Here, we chose $y = 1$, but it could have been anything.

$$x + y = 3$$

x	0	3	2
y	3	0	1

$$5x + 8y = 21$$

x	0	$\frac{21}{5} = 4.2$	$\frac{13}{5} = 2.6$
y	$\frac{21}{8} = 2.6$	0	1

The graph shows that the two lines cross at $x = 1$, $y = 2$, which was the value we found in Answer 5.19.

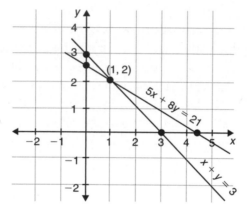

What if we are presented with a graph and are asked to pick the system of equations it represents?

Example 5.22.

Let's use the same graph as above, from Example 5.19, and choose which of the following is the pair of simultaneous equations for that graph.

A. $x + y = 3$

$5x + 8y = -21$

B. $x + y = 3$

$5x - 8y = 21$

C. $x + y = 3$

$5x + 8y = 21$

D. $x - y = 3$

$5x - 8y = 21$

Answer 5.22.

(C) $x + y = 3$, $5x + 8y = 21$. We can read the solution from the graph as the point $(1, 2)$, or $x = 1$, $y = 2$. Just looking at the first equations in each of the answer choices eliminates (D). So it remains to check whether $x = 1$, $y = 2$ satisfies the second equation in (A), (B), and (C). It doesn't work for (A) or (B), so the answer must be (C). But just to be sure, we have to check the second equation in (C), and it works.

Inequalities

If, instead of the equals sign, we wanted an "unequals" sign. We have a choice of five signs, listed below. The a and b can stand for numbers or variables. On the number line (see Chapter 2), anything to the left of a value is "less than" that value, and anything to the right of a value is "greater than" that value.

1. $a < b$ a "is less than" b. So if $a < 7$, a can be any number less than 7, such as -3, 0, 5, 6, but not 7 or anything greater than 7.

2. $a > b$ a "is greater than" b. So if $a > 4$, a can be 5, 7, 100, or any number more than 4, but not 4 or anything less than 4.

3. $a \le b$ a "is less than or equal to" b. So if $a \le 3 + 7$, then a is any number less than 10, *including* 10.

4. $a \ge b$ a "is greater than or equal to" b. So if $a \ge 6 - 2$, then a is any number greater than 4, *including* 4.

5. $a \ne b$ a and b aren't equal (it doesn't say which is larger). So if a is 5, then b cannot be 5.

Solving Linear Inequalities

Inequalities can be combined to indicate a specific range of values. For example, $-2 \le x < 6$, with x being a whole number, means x can be -2, -1, 0, 1, 2, 3, 4, or 5. A set of numbers is sometimes written with braces, such as $\{-2, -1, 0, 1, 2, 3, 4, 5\}$. Notice that it includes -2 here but not 6.

Working with inequalities is similar to working with equalities—whatever we do to one side, we must also do to the other side of the inequality. The only difference is that if we multiply or divide an inequality by a negative, the inequality sign switches. This is due to the hierarchy of negative numbers: although 5 is less than 7, -5 is greater than -7.

HINT

If we multiply or divide an inequality by a negative, the inequality sign switches. To help you remember this, just think of the number line, where $-10 < -5$, but $10 > 5$. If we multiply (or divide) $-10 < -5$ by -1, we get $10 > 5$. The inequality sign switches.

Inequalities are used if there is a limit. For example, if you have only $10 to spend at a store, your purchases (plus tax) must be $\leq$ $10.

Example 5.23.

Jamal is on commission and he wants to earn at least $1,000 in a certain amount of time. His commission is $40 per sale. How many sales does he have to make to meet his goal?

Answer 5.23.

The inequality to use is sales $\times$ commission $\geq$ $1,000. In this case, let's let sales $= x$, and the inequality becomes $40x \geq 1000$. Dividing by 40 on both sides of the inequality, $x \geq 25$. Jamal must make at least 25 sales. Notice how the symbol $\geq$, "greater than or equal to," translates into "at least" in plain English.

Graphing Linear Inequalities

To solve linear inequalities graphically, follow the same procedure as above for linear equalities and plot the line as though the inequality were an equals sign.

1. Write the inequality as an equality, and put it in slope-intercept form.

2. Graph the equality (as we did in Example 5.15). If it is included (that is, if the sign includes equals, as in $\leq$ or $\geq$), graph it as a solid line; if not, make it a dashed line.

3. Determine which points are included in the inequality, and shade that portion of the graph.

4. As a check of whether the correct side of the graph is shaded, pick a point in the shaded portion and see whether the inequality is true for it. A quick way to check is to substitute the origin ($x = 0$, $y = 0$) into the inequality. If the origin is in the shaded portion and the quick check is true, you have shaded the correct side. If the origin is in the nonshaded portion, and the quick check is false, you also have shaded the correct side.

Example 5.24.

Show $y \geq x - 2$ graphically.

Answer 5.24.

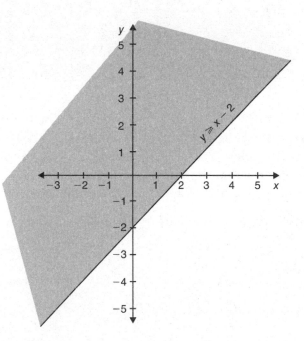

To check whether we shaded the correct side of the line, we pick the origin, which is in the shaded portion. Indeed, when we substitute $x = 0$ and $y = 0$ into the inequality, we get $0 \geq -2$, which is true.

Graphing Simultaneous Linear Inequalities

Now that we know how to graph a linear inequality, what about simultaneous linear inequalities? This is rather straightforward: graph each inequality individually, shading the side as determined above, and the solution to both inequalities is the area where the shading overlaps. This is shown by the following graph.

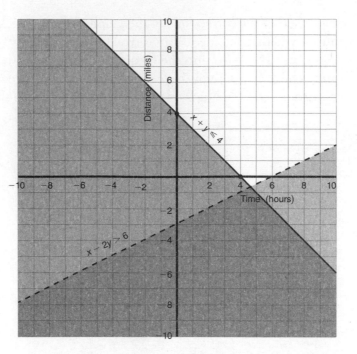

The straight lines are $x + y \le 4$ and $x - 2y > 6$. The double-shaded portion, including the solid line but excluding the dashed line, contains all the points that are solutions to both inequalities. Just in case you are wondering why the area under $x - 2y > 6$ is shaded when it has a $>$ sign, remember two facts: the equation should be rewritten in point slope form ($y = mx + b$), and this involves dividing the inequality by -2, but when we multiply or divide an inequality by a negative number, the inequality switches. So $x - 2y > 6$ actually turns out to be $y < -\dfrac{1}{2}x - 3$.

Exercises

1. Simplify by combining like terms: $10x^3 + 5x^2 - 2x + 1 - x^3 - 10x^2 - 2x - 1$.

    A. $-x^3 + 5x^2 - 4x$

    B. $9x^3 - 5x^2 - 4x$

    C. $9x^3 + 5x^2 - 4x$

    D. $9x^3 - 5x^2 - 4x - 1$

2. How much smaller is $x^2 - 3$ than $2x + 1$?

    A. $x^2 - 2x - 4$

    B. $x^2 + 2x - 2$

    C. $-x^2 + 2x + 4$

    D. $-x^2 + 2x + 2$

3. Simplify: $x(x - 3) - (x^2 - 6) - (x^2 - 2x + 9)$.

    A. $-(x^2 + x + 3)$

    B. $x^2 + x + 3$

    C. $x^2 - x - 15$

    D. $-x^2 - 5x + 3$

4. If $x = -3$, $y = 2$, and $z = 0$, the value of $xy + 8z - x$ is $\boxed{}$.

5. The area of a trapezoid is given by $A = \dfrac{1}{2}h(b_1 + b_2)$. What is the area when $h - 1$, $b_1 - 14$, and $b_2 = 8$?

    A. 22

    B. 11

    C. 3

    D. -3

6. What is the product of $3x^2y(x^2 + xy - y^2)$?

    A. $3x^4y + 3x^3y^2 + 3x^2y^3$

    B. $3x^4y + 3x^3y^2 - 3x^2y^3$

    C. $3x^4y + x^3y^2 - x^2y^3$

    D. $3x^5$

7. If $10x + 1 = 8x + 2$, $x = \boxed{}$.

    A. $\dfrac{1}{2}$

    B. 2

    C. $\dfrac{2}{3}$

    D. 1

8. The sum of a certain number and 6 is 20. Half that number is $\boxed{}$.

9. The number of men in an adult education course is one less than three times the number of women. If 39 people are taking the course, how many are women?

    A. 29

    B. 10

    C. 20

    D. cannot tell from the data given

10. Zeb and his friend Jon cleaned a yard for $80 total. If Zeb worked three times as many hours as Jon, how much of the money should Zeb get?

    A. $20

    B. $40

    C. $60

    D. $80

11. A seesaw is balanced when the force (weight) at one end times its distance from the fulcrum (pivot) equals the weight at the other end times its distance from the fulcrum. Marianne, who weighs 48 pounds, is sitting 5 feet from the pivot of a 9-foot seesaw. Her friend Jimmy joins her, but when she gets on, the seesaw is perfectly balanced in the air. How much does Jimmy weigh?

    A. 48 pounds

    B. 60 pounds

    C. 38.4 pounds

    D. 90 pounds

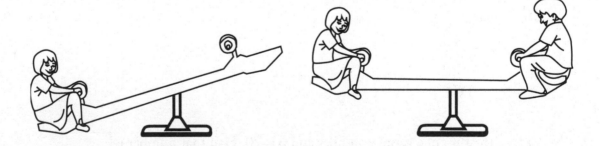

12. Write the formula for the following table by inserting the correct parts of the formula from the choices given into the blank boxes.

x	0	3	5	7	9
y	2	5	7	9	11

$y = $ ☐ ☐ ☐ .

| 2 | 3 | + | x | − | 2x | 3x |

13. The formula for the following graph is

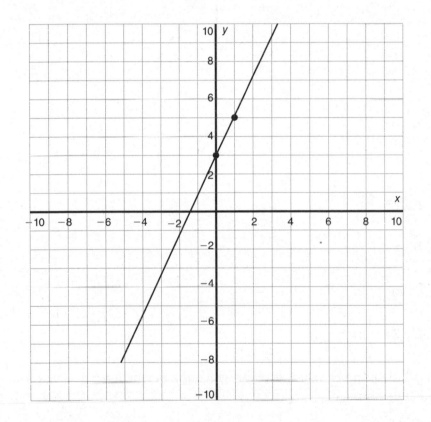

A. $y = 2x + 3$

B. $2y = 4x + 6$

C. $2x = y - 3$

D. All of these

14. Which of these equations is perpendicular to the line in the following graph?

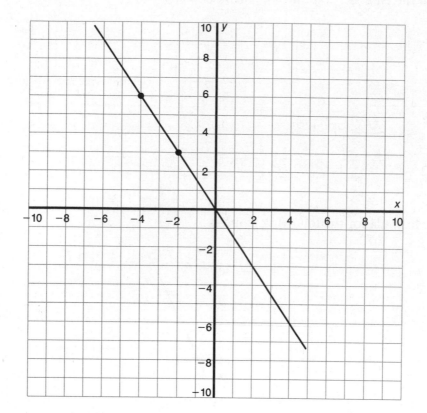

A. $y = \dfrac{x+5}{3}$

B. $y = \dfrac{x+5}{2}$

C. $3y = 2x + 7$

D. $2y = 3x + \dfrac{5}{2}$

15. The whole number values of x for which the inequality $-3 \le 3(x + 2) < 2$ is true are the whole numbers

A. 2 and 3

B. -2 to 3

C. -3 to 2

D. -3 and -2

Solutions

Answer 1. (B) $9x^3 - 5x^2 - 4x$. Combine only like terms, and keep the signs straight. As you determine what each term will be, you can eliminate some answer choices. For example, the first term will be $9x^3$, so answer choice (A) is eliminated. You may be able to solve this problem by elimination without figuring out each term.

Answer 2. (C) $-x^2 + 2x + 4$. "How much smaller" means subtract. If you are unsure which expression to subtract from which, substitute some numbers into the sentence: "How much smaller is 6 than 8?" That will remind you to subtract the first expression from the second, so the calculation is

$$(2x + 1) - (x^2 - 3) = 2x + 1 - x^2 + 3 = -x^2 + 2x + 4$$

Answer 3. (A) $-(x^2 + x + 3)$. The actual answer is $-x^2 - x - 3$, but this simplifies to answer choice (A). The other answer choices are wrong because they didn't carry the minus signs through all of the terms in parentheses.

Answer 4. -3. Remember that multiplying anything by 0 equals 0 and that minus a negative number is a positive number.

Answer 5. (B) 11. Answer choice (D) can be eliminated right away because areas cannot be negative. The other two answer choices are due to calculation errors, such as not dividing by 2.

Answer 6. (B) $3x^4y + 3x^3y^2 - 3x^2y^3$. Answer choice (D) should be eliminated right away because it doesn't have three terms. Answer choices (A) and (C) are due to calculation errors, such as an error with a minus sign.

Answer 7. $\frac{1}{2}$. Do the algebra without making mistakes. This problem can also be answered by substituting the values for x into the equation.

Answer 8. 7. Find x from the equation $x + 6 = 20$, and then take half of the solution.

Answer 9. (B) 10. This problem involves two equations with two unknowns, say, $M =$ number of men and $W =$ number of women. If the number of men is one less than three times the number of women, $M = 3W - 1$. The second equation shows that there are 39 people in the class, so $M + W = 39$. Solve the simultaneous equations by substituting $M = 3W - 1$ into $M + W = 39$, getting $(3W - 1) + W = 39$, or $4W = 40$, so $W = 10$. The answer can also be found by substituting the answer choices into the given information.

Answer 10. (C) $60. The easiest way to find the answer here is by eliminating answers. Answer choices (A) and (B) should be eliminated right away because they wouldn't give Zeb more money than Jon. Eliminate answer choice (D) because that would give Zeb all of the money. To work it out algebraically, if Jon worked x hours, Zeb worked $3x$ hours, and thus $x + 3x = \$80$, or $x = \$20$. Zeb's share is $3x = \$60$. This problem can be done in several other ways, such as using proportions: $\dfrac{\text{John}}{\text{Zeb}} = \dfrac{1}{3} = \dfrac{80 - x}{x}$, which gives $x = 240 - 3x$, or $x = 60$.

Answer 11. (B) 60 pounds. When the seesaw is perfectly balanced, Marianne's weight times her distance from the pivot equals Janey's weight times her distance from the pivot. So the equation, where x is Janey's weight, is

$$48(5) = x(9 - 5)$$

$$240 = 4x$$

$$x = 60 \text{ pounds}$$

Answer 12. $y = x + 2$. From the table, the y-intercept is 2 (when $x = 0$). The slope, using the formula $m = \dfrac{y_2 - y_1}{x_2 - x_1}$ (which is on the GED® test formula sheet), is $m = \dfrac{2}{2} = 1$. Thus, the equation in slope-intercept form is $y = x + 2$.

Answer 13. (D) All of these. The equations are all the same line. Answer choice (B) is just twice answer choice (A), and answer choice (C) is just another way of writing answer choice (A).

Answer 14. (C) $3y = 2x + 7$. The slope of the line in the graph is $\dfrac{3}{-2} = -\dfrac{3}{2}$, so a line perpendicular to it would have a negative reciprocal slope, or $\dfrac{2}{3}$. The only answer choice with a slope of $\dfrac{2}{3}$ is (C).

Answer 15. (D) -3 and -2. The inequality yields the following boundaries for x:

$$-3 \leq 3(x + 2) < 2$$

Remove parentheses:	$-3 \leq 3x + 6 < 2$
Subtract 6 from all parts	$-9 \leq 3x < -4$
Divide by 3, the coefficient of x	$-3 \leq x < -\dfrac{4}{3}$

The whole numbers in that interval are $-3, -2$.

Chapter 6

The X (Squared) Factor

Polynomials

Mathematical expressions with just two terms, such as $x + 7$ or $a + b$, are called **binomials**. (bi = "two," as in $bicycle$), and those with three terms are called **trinomials** (tri = "three," as in $tricycle$). But all expressions of two or more terms are called **polynomials** ($poly$ = "many"). A **monomial** is the name for an expression with only one term ($mono$ = "one," as in $monorail$).

Adding and Subtracting Polynomials

We already worked with adding and subtracting polynomials in Chapter 5. They involve combining like terms—those with unknowns that are identical. For example,

$$(3x^2 + 2x + 4) + (x^2 - x - 7) = 4x^2 + x - 3.$$

Multiplying Polynomials

We also already saw multiplication of a polynomial by a monomial in Chapter 5. But when we multiply a polynomial by a *polynomial*, things get a little more complicated. The multiplication that is most often encountered is the multiplication of two binomials, and that is what we will concentrate on here because it has a lot of uses and implications.

An example of the multiplication of two binomials is $(x + 2)(x - 4)$. One way to do this is by longhand—multiply $(x - 4)$ by x and then multiply $(x - 4)$ by 2 and combine like terms. But there is a better method, called the F-O-I-L method. F-O-I-L is a mnemonic, a word to remember something. (We encountered the mnemonic PEMDAS in Chapter 4 when we talked about the order of operations.)

The letters in F-O-I-L stand for

First-Outside-Inside-Last.

This means that, when we multiply two binomials, we multiply the first terms, then the outside terms, then the inside terms, and finally the last terms. It's worth learning the F-O-I-L method because it makes working with binomials so much easier than multiplying by brute force.

The following explanation of F-O-I-L uses specific binomials because it is easier to understand that way, but F-O-I-L works with any coefficients and any constants. (Many books use ax, b, cx, and d instead of numbers, but that just confuses everything.) Let's use the binomials $(x + 6)(2x + 3)$:

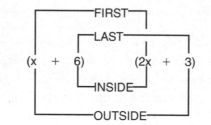

The *first* terms are x and $2x$, the *outside* terms are x and 3, the *inside* terms are 6 and $2x$, and the *last* terms are 6 and 3.

Now, what do we do with them? We multiply each of them, in F-O-I-L order, combine like terms, and we end up with the product when we multiply two binomials. It really is the same as the longhand method, but it's more straightforward.

Here, when we multiply the first terms, we get

$$+ 2x^2,$$

when we multiply the outside terms, we get

$$+ 3x$$

when we multiply the inside terms, we get

$$+ 12x,$$

and when we multiply the last terms, we get

$$+ 18.$$

Then when we add it all up, we get the product:

$$2x^2 + 3x + 12x + 18,$$

which equals

$$2x^2 + 15x + 18.$$

This trinomial has a specific name; it is called a **quadratic**, which essentially is an expression in which the highest power of the unknown is 2. We talk more about quadratics later in this chapter.

Let's look at a few examples of multiplying two binomials.

Example 6.1.

Multiply $(x + 3)(3x + 1)$.

Answer 6.1.

$3x^2 + 10x + 3$. The F-O-I-L method gives us $3x^2 + x + 9x + 3 = 3x^2 + 10x + 3$.

By adding the outside and inside products together right away (since they are like terms), with a little practice you can do such multiplication in your head.

Specifically let's look at how the minus signs, if there are any, are handled—they go with their coefficient or constant. And we remember that when we multiply two like signs, the product is positive (even two minuses give a positive), and when we multiply two unlike signs, the product is negative.

Example 6.2.

Multiply $(2x - 2)(x + 5)$.

$2x^2 + 8x - 10$. Are you following how we got this answer? The $8x$ term comes from the sum of the outside ($10x$) and inside ($-2x$) products.

Example 6.3.

Multiply $(x + 6)(x - 8)$.

Answer 6.3.

$x^2 - 2x - 48$. Is it getting any easier? The $-2x$ term comes from the sum of $-8x$ and $+6x$.

Example 6.4.

Multiply $(2x - 3)(4x - 1)$.

Answer 6.4.

$8x^2 - 14x + 3$. Even though you probably got this one, let's go through it because of all the minus signs. The $8x^2$ is no problem, the product of the first terms. The $-14x$ comes from the sum of the outside ($-2x$) and inside ($-12x$) products. The $+3$ comes from the product of the last terms $(-3)(-1)$. Again, remember that a minus times a minus (like signs) is always a positive.

Dividing Polynomials

As we saw with numbers in Chapters 2 and 3, if we divide a sum by a number, it has to be a factor of every term or we cannot do the division. For example, $\dfrac{50 + 42 - 30}{2} = 25 + 21 - 15$ because 2 divides into each term evenly, but we can't do something similar with $\dfrac{49 + 42 - 31}{2}$ because 2 doesn't divide evenly into 49 or 31. Yes, we could have just added the numbers and divided the sum by 2, easy enough, but we're trying to make a point here.

And the point is that, if we now have a polynomial divided by anything, it had better divide evenly into every term of the polynomial or all (or most) bets are off. For example,

$$\frac{15x^3 - 40x^2 - 15}{-5} = -3x^3 + 8x^2 + 3$$

because -5 divides evenly into each term in the numerator. If the coefficient of the x^3 term had been a 7, then we wouldn't have been able to do this division so smoothly. By the way, the reason we say most (and not all) bets are off is that it can be done, but we'll leave that to the mathematicians.

Just as factoring was important when dividing any two quantities, whether they are numbers, fractions, or anything else, factoring is an important part of dividing two polynomials. For example, by recognizing that there is a common factor (x) in the numerator and denominator of $\dfrac{x^3 + 6x^2 + 5x}{x^2 + x}$, we get

$$\frac{x^3 + 6x^2 + 5x}{x^2 + x} = \frac{x(x^2 + 6x + 5)}{x(x+1)} = \frac{x^2 + 6x + 5}{x+1}.$$

In fact, in this example, the numerator can be factored further, and the whole expression equals $(x + 5)$. But we are getting ahead of ourselves a little here. First we must talk about quadratic equations and how to factor them, the topic of the next section.

Quadratic Equations

A **quadratic equation** is an equation in which the unknown is squared and there is no higher power of the unknown. It is okay if there are no lower powers of the unknown (in other words, no "x" term or no "pure number" term). An example of a quadratic equation is $x^2 + x - 6 = 0$, and so are $x^2 - 9 = 0$ and $x^2 + 3x = 0$. Quadratic equations always have two answers for the value of the unknown (even though at times they are the same number twice).

The general form of a quadratic equation is $ax^2 + bx + c = 0$, where b and c can be any numbers, even 0, as we saw with the examples of $x^2 - 9 = 0$ and $x^2 + 3x = 0$. If $a = 0$, though, we no longer have a quadratic—according to the definition, there has to be a squared term.

Solving a Quadratic Equation

The solutions to a quadratic equation are based on a simple fact: if two factors are multiplied together and the product is 0, then either one or both of the factors must equal 0. There just are no two nonzero numbers whose product is 0. Period. We always write a quadratic equation on one side of the equals sign with 0 on the other so we can see what the factors are.

In other words, if we can find two factors for a quadratic equation, its solution is found by setting each of the factors equal to zero and solving for the unknown variable. Each of the factors will contain the unknown to only the first power (x), and we already know how to solve that kind of equation. Then we end up with two **roots** (solutions) for the quadratic equation.

There are many ways to solve a quadratic equation by factoring, and they mostly depend on recognizing what kind of equation is presented and what would be easiest.

1. *Difference of two squares.* If we can recognize the quadratic equation as the difference of two squares (a perfect square, a minus sign, and another perfect square, with no x term, such as $x^2 - 25$), the factors are the sum and difference of the square roots, here $(x - 5)(x + 5)$. Setting each factor equal to 0 we get, $x = +5$ or -5, written as $x = \pm 5$ because $\pm$ means plus or minus.

   Another example of the difference of two squares is the quadratic equation $x^2 - 9 = 0$. The factors are $(x + 3)$ and $(x - 3)$. If we set each factor equal to 0, the two solutions to $x^2 - 9 = 0$ are $x = \pm 3$.

   Yet another example of the difference of two squares actually is the difference of two quantities to the fourth power, such as $x^4 - 16 = 0$. If we look at this as the difference of two squares, then the factors are the sum and difference of the square roots, $(x^2 + 4)$ and $(x^2 - 4)$, but the last factor is itself the difference of two squares, so the factors of $x^4 - 16$ are $(x^2 + 4)(x + 2)(x - 2)$. In this case, we know two of the roots are $x = \pm 2$, but $x^2 + 4$ doesn't have any real roots because there isn't any number that, when squared, will yield -4. Such roots are called **imaginary**, and they are beyond the scope of this book. We will work only with real numbers.

2. *Factoring out a common factor.* This means recognizing that there is a common factor to begin with. For equations without the constant (c) term, the factor is x, as in $x^2 + 3x = 0$. If the x is factored out, we get $x(x + 3)$ as the two factors, and setting each equal to 0 here gives us $x = 0$ or -3 as the roots.

3. *Factoring out binomials.* This method involves the reverse of what we just learned about multiplying two binomials to get a quadratic equation. Some of you may already know how to do this. For a quadratic equation of the general form, $ax^2 + bx + c = 0$, where $a = 1$, this involves finding two numbers whose product is c, the last term of the quadratic equation, and whose sum is b, the coefficient of the x term. However, the quadratic formula, discussed next, works for all binomial factoring and doesn't involve the guesswork of this method.

HINT

   All quadratic equations can be solved by using the quadratic formula. This formula is on the formula sheet of the GED® test. Although it looks complicated, it is the go-to way to solve quadratics.

4. *Use the quadratic formula.* Even though you don't have to memorize this formula, you do have to know how to use it. First, the quadratic equation has to be in the general quadratic equation form of $ax^2 + bx + c = 0$, which may involve combining like

terms. Once it is in this form, we need only to determine the values of a, b, and c (they are the coefficients), and plug them into the formula, which is

$$x = \frac{-b \pm \sqrt{b^2 - 4ac}}{2a}.$$

As we said before, the value of a cannot be zero. There are two reasons for this: (1) if $a = 0$, the equation would be linear and not quadratic, of the form $bx + c = 0$; and (2) division by 0 is undefined.

Notice that there is a $\pm$ sign in the equation, which gives the two roots that all quadratic equations have. Also, the equation involves a square root. The value under the square root sign is called the **discriminant**. On the GED® test, the discriminant will usually be a common square, such as 16 or 25, but it can also be 7. Remember that you don't have to know how to find a square root for the GED® test (other than on the calculator), but you should be able to recognize common squares and square roots (see Chapter 4).

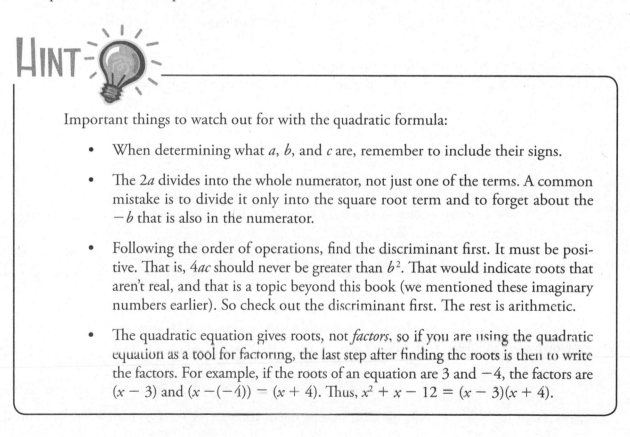

HINT

Important things to watch out for with the quadratic formula:

- When determining what a, b, and c are, remember to include their signs.

- The $2a$ divides into the whole numerator, not just one of the terms. A common mistake is to divide it only into the square root term and to forget about the $-b$ that is also in the numerator.

- Following the order of operations, find the discriminant first. It must be positive. That is, $4ac$ should never be greater than b^2. That would indicate roots that aren't real, and that is a topic beyond this book (we mentioned these imaginary numbers earlier). So check out the discriminant first. The rest is arithmetic.

- The quadratic equation gives roots, not *factors*, so if you are using the quadratic equation as a tool for factoring, the last step after finding the roots is then to write the factors. For example, if the roots of an equation are 3 and -4, the factors are $(x - 3)$ and $(x - (-4)) = (x + 4)$. Thus, $x^2 + x - 12 = (x - 3)(x + 4)$.

Example 6.5.

Find the roots of $x^2 + x - 6 = 0$ by using the quadratic formula.

Answer 6.5.

2 or -3. The equation is in the correct form, with $a = 1$, $b = 1$, and $c = -6$. Remember that the minus signs go with the coefficients; actually, so do the plus signs.

So the first thing to do is to calculate the discriminant, $b^2 - 4ac$, to make sure it is ≥ 0. Otherwise, there is no real solution for the quadratic. Here, it is $1^2 - 4(1)(-6) = 1 + 24 = 25$. Then just plug the values into the quadratic formula to get the two answers for x.

$$x = \frac{-b \pm \sqrt{b^2 - 4ac}}{2a} = \frac{-(1) \pm \sqrt{25}}{2(1)} = \frac{-1 \pm 5}{2} = \frac{-1 + 5}{2} = 2, \text{ or } \frac{-1 - 5}{2} = -3.$$

Sure enough, if we substitute 2 or -3 into the original equation, it equals 0 in both cases.

As we said, the quadratic formula works for all quadratic equations, even the ones used as examples for the difference of two squares and factoring out a common factor.

1. For $x^2 - 25 = 0$, with $a = 1$, $b = 0$, $c = -25$. First check $b^2 - 4ac = 0 - 4(1)(-25)$ $= 100$, so the roots are real.

$$x = \frac{-b \pm \sqrt{b^2 - 4ac}}{2a} = \pm \frac{\sqrt{100}}{2} = \pm \frac{10}{2} = \pm 5.$$

2. For $x^2 + 3x = 0$, $a = 1$, $b = 3$, and $c = 0$. First check $b^2 - 4ac = 3^2 - 4(1)(0) = 9$, so the roots are real.

$$x = \frac{-b \pm \sqrt{b^2 - 4ac}}{2a} = \frac{-3 \pm \sqrt{9}}{2} = \frac{-3 + 3}{2} = 0, \text{ or } \frac{-3 - 3}{2} = -3.$$

We thus see that the quadratic formula will work for all quadratic equations, and there are two real answers if the discriminant is ≥ 0.

Can you see that if the discriminant $= 0$ (that is, if $b^2 = 4ac$), then there are two identical answers equal to $\frac{-b}{2a}$? An example of this case is the quadratic $x^2 + 6x + 9$, with both roots equaling $\frac{-b \pm \sqrt{0}}{2a} = \frac{-6}{2} = -3$. The factors of $x^2 + 6x + 9$ would then be $(x + 3)(x + 3)$.

Example 6.6.

What are the roots of $x^2 - 10x + 25 = 0$?

Answer 6.6.

For $x^2 - 10x + 25 = 0$, $a = 1$, $b = -10$, and $c = 25$. The discriminant is $b^2 - 4ac = (-10)^2 - 4(1)(25) = 100 - 100 = 0$. Thus,

$$x = \frac{-b}{2a} = \frac{10}{2(1)} = 5$$

So the two factors of $x^2 - 10x + 25$ are $(x - 5)(x - 5)$.

Checking the Roots

We can check the roots by substituting the values found for x into the original equations.

For $x^2 - 9 = 0$, we found $x = \pm 3$. Substituting these values into the equation gives us $(3)^2 - 9 = 0$ and $(-3)^2 - 9 = 0$. So both roots check.

For $x^2 + 3x = 0$, the roots are $x = 0, -3$. Using substitution for $x = 0$, $(0)^2 + 3(0) = 0 + 0 = 0$, and for -3, $(-3)^2 + 3(-3) = 9 - 9 = 0$. Both roots check.

For $x^2 - 10x + 25 = 0$, we found a double root of 5, and $(5)^2 - 10(5) + 25 = 25 - 50 + 25 = 0$. It checks, too.

Example 6.7.

Factor $x^2 + 6x + 8$.

Answer 6.7.

$(x + 2)(x + 4)$. Use the quadratic formula to find the roots of $x^2 + 6x + 8 = 0$, and then convert the information about the roots into the factors. Here, $a = 1$, $b = 6$, and $c = 8$. The discriminant is $b^2 - 4ac = 36 - 32 = 4$. It is positive, so we get

$$x = \frac{-b \pm \sqrt{b^2 - 4ac}}{2a} = \frac{-6 \pm \sqrt{4}}{2} = \frac{-6 \pm 2}{2} = -2 \text{ or } -4.$$

Then the factors are $x^2 + 6x + 8 = (x - (-2))(x - (-4)) = (x + 2)(x + 4)$.

Example 6.8.

Simplify $\dfrac{x^2+6x+8}{x^2+5x+6}$.

Answer 6.8.

$\dfrac{x+4}{x+3}$. Factor the numerator and denominator to get $\dfrac{(x+2)(x+4)}{(x+2)(x+3)}=\dfrac{x+4}{x+3}$ by canceling out the $(x + 2)$ in the numerator and denominator. We get the factors of the numerator from Example 6.7. The factors of the denominator come from a similar method:

$$x = \frac{-b \pm \sqrt{b^2-4ac}}{2a} = \frac{-5 \pm \sqrt{1}}{2} = \frac{-5 \pm 1}{2} = -2 \text{ and } -3, \text{ so the factors are } (x+2)(x+3).$$

Functions

Many equations can be classified as **functions**, which have a special relationship between input values and output values. Specifically, for each input value, there can be only one output value.

HINT

Think of a function as a machine into which you feed an x and it gives you an $f(x)$. Each x gets just one $f(x)$, but $f(x)$ can be an output for many x's.

Note that $f(x)$ means "function of x," not f multiplied by x.

The quadratic equation $y = ax^2 + bx + c$ is a function. When $y = 0$ (as we showed in Chapter 5, this is when the function crosses the x-axis), the equation is called a quadratic equation with **roots** (answers or solutions) at those x values.

Instead of using y, functions are usually indicated by $F(x)$ or $f(x)$, which are interpreted as "function of x," where x is a variable. Functions (and, of course, variables) can be indicated by any letter, however. The idea is the same.

We actually encountered a function when we graphed the linear equation $y = x - 2$, which also could have been written as $f(x) = x - 2$. For a relation to be a function, each value of x, the **domain**, will give one (and only one) value for y, the **range**. So for $f(x) = x - 2$, if $x = 1$, $f(x) = -2$ and cannot also equal another value.

HINT

Sometimes it is difficult to remember which is the domain and which is the range. Try this: *x* comes before *y* in the alphabet, and *domain* comes before *range* in the alphabet, so *x* is the domain and *y* is the range.

Functions do not have to be straight lines, however. In fact, graphs of functions often are curves of one type or another. Quadratic equations have a distinct shape when graphed, called a **parabola**, kind of like a ∪ or ∩ with its sides flared out.

The path of a baseball when it is hit by a bat is a perfect example of a quadratic function. If you think of the ground as *x* and the height of the ball above the ground as *y*, this is obviously a function. At any place on the field (*x*), the ball is at a certain height (*y*)—outfielders just kind of know (from experience) where to stand to catch the ball. An outfielder won't attempt to catch a ball where he is standing at the time the ball is hit if he knows it will be 20 feet above him there—he backs up to where he thinks the ball will be at a catchable height. Other examples of parabolas are the path of the bullet when fired from a gun and the arch made by a basketball in a free throw.

The other way around is okay—for each *y* value there can be more than one *x* value. Certainly, when the batter hits the ball, it is at about the same height as when the outfielder catches it. So when the ball is 4 feet off the ground ($f(x) = 4$, or $y = 4$), it is over home plate and then later it is in the outfield, some 300 or so feet away from the batter. In fact, the outfielder is calculating (not with a calculator but with instinct) that the path of the ball up to its highest point (called the **vertex**) will make it "catchable" at the place on the field that is a match to how far from that vertex the ball was when it was hit.

Incidentally, a ⊂-shaped or ⊃-shaped graph is not a function because for any *x* there are two values for *y*.

Evaluating Functions

Functions can be evaluated. In fact, for many of the substitution problems we have already encountered, we have actually evaluated a function. As an example, to evaluate $x^2 + 3x - 5$ when $x = 2$, which means to substitute 2 for each *x*, we calculate $(2)^2 + 3(2) - 5 = 4 + 6 - 5 = 5$. This is exactly the same as the following problem: "$f(x) = x^2 + 3x - 5$, find $f(2)$." It simply means "replace all the *x*'s with 2's."

Taking this replacement one step further, we can also evaluate the following. If $f(x) = 2x - 4$, what is $f(y)$? We just have to remember that the method doesn't change—we replace every x by y, so $f(y) = 2y - 4$. What then would $f(x + 1)$ be? Again, the method doesn't change—replace each x with $(x + 1)$ to get $f(x + 1) = 2(x + 1) - 4$, which we would then simplify to $f(x + 1) = 2x + 2 - 4 = 2x - 2$. Don't let the reuse of x throw you; just follow the rule.

Graphing Functions

Graphing a function is the same as the method we used before—just think of $f(x)$ as y. Calculate three points if it is a linear (first-degree) equation, and several if it is a quadratic (second-degree) equation, and then connect the points. Again, if it is a function, then each x value will have only one $f(x)$ value. Likewise, if you are given a graph, you can tell right away if it is a function by checking whether the graphed curve doubles back on itself as you look at it above the x-axis. And then we can even match a graph to its function (at least on a multiple-choice question) by evaluating $f(x)$ for some x points and seeing whether they match the graph.

Example 6.9.

For which function $f(x)$ is the following graph a possibility?

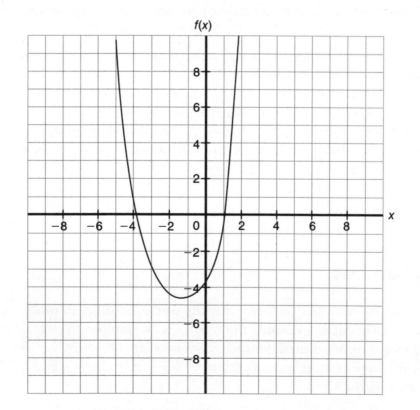

A. $f(x) = 2x - 4$

B. $f(x) = x^2 + 3x - 4$

C. $f(x) = x^3 + 2x - 4$

D. $f(x) = x^2 + 8x - 8$

Answer 6.9.

(B) $f(x) = x^2 + 3x - 4$. We can eliminate (A) because that is the equation of a line, not a curve, and (C) because that is a cubic (third-degree) equation, not a quadratic equation. Then the task is to check out some points on the graphs of (B) and (D) because they are both quadratics. The first point to check is always $f(0)$ because it is the easiest. For answer choice (B), $f(0) = -4$; for answer choice (D), $f(0) = -8$. We need to go no further: the graph is clearly not choice (D), and the answer is (B).

Describing Graphs

We already talked about parabolas, the ∪-shaped curves of quadratic equations. It is a function because for every x there is one and only one y. Another function is an **exponential** function, which is typically a slow growth and then an accelerated growth, or a slow decay and then an accelerated decay. A typical equation for an exponential function is $f(x) = C^x$, where the x is the exponent, and C can be any constant. The graph for $f(x) = 2^x$ is shown below.

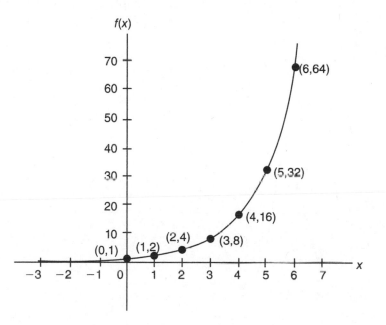

An example of exponential growth is bacterial growth. As a simple example, if you had a bacterial infection and the number of bacteria doubled every hour, at the end of a day there would be $2^{24} = 16,777,216$, or more than 16 million times the number of bacteria. No wonder you can feel sick so suddenly. Now let's suppose at the end of the day, you take an antibiotic that kills half the bacteria every hour. At the end of the next day (assuming the bacteria didn't grow any more), the 2^{24} bacteria would be dead, reduced by a factor of $\left(\dfrac{1}{2}\right)^{24} = \dfrac{1}{16,777,216}$. Take your medicine—it works!

Another type of graph displays several features of interest. You should be able to recognize these features on a graph just by their names. First of all, note that this graph, although it seems to be several graphs together, is a function called a **composite function**. There are no places where a value of x gives more than one value of y. It illustrates that a function does not have to be just one particular graph.

- Point A is an **endpoint**, and so is point G. They are the ends of the graph as drawn.

- Point B is a **relative maximum**, meaning it is the highest point relative to the points on either side of it. So is point E, and in addition, point E can be called an absolute maximum because it is the highest point on the whole graph.

- Point C is a **relative minimum**, meaning it is the lowest point relative to the points on either side of it.

- The graph is increasing for the following intervals: AB, CD, and DE.

- The graph is decreasing for the following intervals: BC, EF, and FG.

- Interval BC is symmetric to interval CD, meaning they are mirror images of each other.

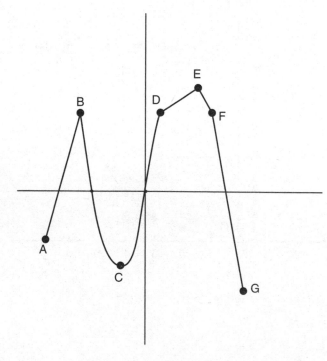

If a function has **periodicity**, there is a repeat of the graph at regular intervals. An example of periodicity would be a steady heartbeat on a heart monitor or the sine function, which involves trigonometry, shown below.

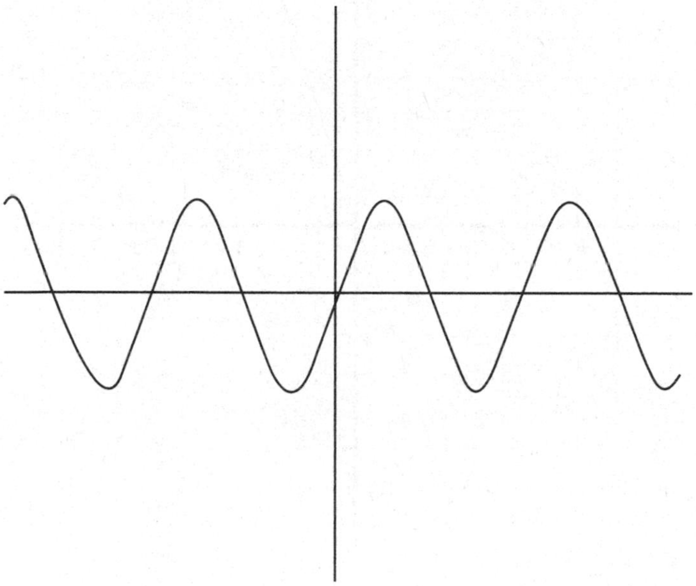

Comparing Functions

Functions come in several forms, as we have seen. For simplicity, let's use the linear function $f(x) = 3x$ as an illustration of these various forms, although all functions, even quadratics, can be presented in these various forms.

- *Algebraic*, or in the form of an equation. Our equation here is function $f(x) = 3x$.

- *Numeric*, or in the form of a table that presents values for x and $f(x)$, such as the one shown below.

x	0	1	2	3
f(x)	0	3	6	9

- *Graphic*, or, as the name states, as a graph, such as the one shown on the next page.

- *Verbal*, or in words, such as "Jenny's distance, walking at a pace of 3 miles an hour."

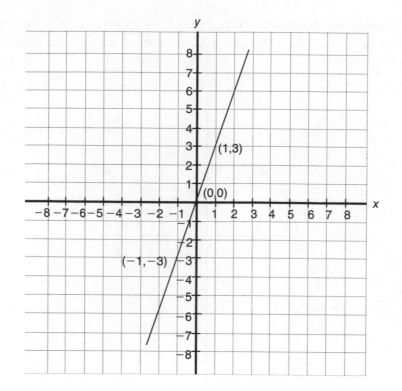

Each of the forms presented here is actually saying the same thing. The verbal sentence can be presented algebraically; that equation can be the basis of the table; and when we graph the values in the table, we get the graph. Actually, any representation can be the starting point.

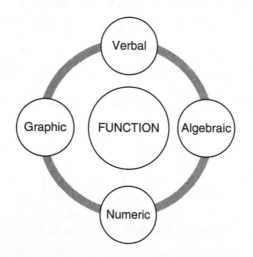

We recognize the coefficient of the *x* term in the equation as the slope, using the $y = mx + b$ slope-intercept form of an equation, and that is shown on the graph. Verbally, the *x* term of the functions is presented here as a rate, and we can get Jenny's distance by using various values for *x* and recording the corresponding $f(x)$ values in a table or on a graph.

Now, let's say that we wanted to compare the rates of two pretty fast animals—a racehorse and a kangaroo—to see which is faster. We could use any of the formats to do that. First, we need some information, such as distance and time to compare the rates of these speedsters. Let's say we know a British racehorse won a 3,200-meter race in a time of 3 minutes and 20 seconds, which is 200 seconds. And a kangaroo is known to be able to cover 2,520 meters in 180 seconds. Which is faster? That was the verbal representation.

The algebraic forms of this same information would be $f(r)_{racehorse} = 3{,}200 = 200r_{racehorse}$ and $f(r)_{kangaroo} = 2{,}520 = 180\,r_{kangaroo}$. A comparison would give us $r_{racehorse} = 16$ meters per second and $r_{kangaroo} = 14$ meters per second. The horse is faster than the kangaroo.

We can even construct tables of distance ($f(x) = d$) versus time ($x = t$) to get a sense of how the two animals compare. The following table shows clearly that the racehorse is faster.

$t_{racehorse}$	$d_{racehorse}$
100 seconds	1,600 meters
150 seconds	2,400 meters
200 seconds	3,200 meters

$t_{kangaroo}$	$d_{kangaroo}$
100 seconds	1,400 meters
150 seconds	2,100 meters
200 seconds	2,800 meters

Graphically, we can compare the slopes (same as rates) of the two graphs generated. The steeper (positive) slope represents the faster rate. So we clearly see that the horse's graph has a steeper slope and thus a faster rate.

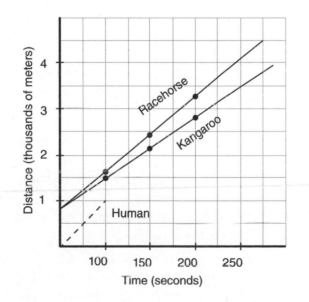

This graph also points out that the axes do not have to be in single units (ones) and can represent, as the distance axis does here, thousands of units. The factor just has to be stated on the graph, such as the "Distance (thousands of meters)" shown on this graph.

Just for fun, the dashed line in the graph represents Olympic times for a human, and this slope is even steeper than a racehorse! But we have to take into account that the record human speed of around 10 seconds for the 100-meter dash can be sustained for only that long—about 10 seconds. That is why the graph is dashed—it is graphed as though that speed can go on for 100 seconds—which is impossible!

As an example of comparing quadratic equations, let's look at the standard equation, $f(x) = ax^2 + bx + c$. The constants a, b, and c determine how the graph will look, but we have to remember that the slope of a quadratic equation is constantly changing (see the figure below), so comparing slopes between two quadratic equations is complicated.

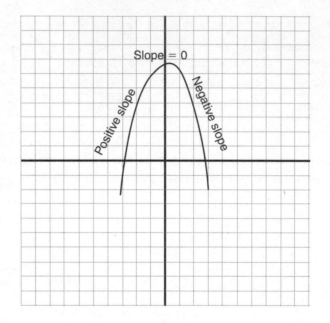

However, we can compare two quadratics in other ways. We mentioned earlier that the path of a baseball hit by the batter is a quadratic function. The a, b, and c constants in the quadratic function are related to how the ball was hit (angle, force, etc.), and that is how we can compare three quadratics that represent the paths of three hit balls. The y-axis represents the height of the ball when it is hit (and caught). The horizontal-axis represents the distance from home plate, which is 0 at the far left, to the outfield fence at the far right.

Curve A shows a pop fly. The curve reaches its peak in the infield and is caught somewhere beyond the base lines. Curve B represents a ball that is hit long and high, but comes down just before the outfield fence, so it is probably a double or a triple (or maybe an inside-the-park home run). Curve C is a home run. The ball comes down over the fence and into the crowd, where some lucky fan catches it and takes it home as a souvenir.

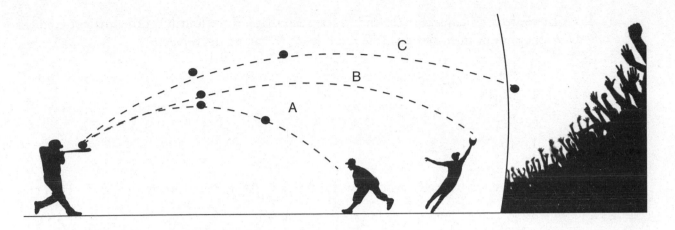

As noted earlier in this chapter, good baseball players get a sense of where the ball will land and place themselves in that position by watching where it reaches its peak (the vertex). Will you ever be able to watch a baseball (or softball) game again without thinking of parabolas?

Exercises

1.   If $x^2 - 18 = 0$, the possible values of x are ⬚ . (Note: Leave answer in terms of the radical sign.)

2.   Solve for x without using a calculator. If $2x^2 = 200$, $x =$

   A. ± 20

   B. 10

   C. ± 10

   D. 100

3.   Solve for x: $3x^2 - 5x - 2 = 0$.

   A. 1 or $\dfrac{2}{3}$

   B. 2 or $-\dfrac{1}{3}$

   C. -1 or $-\dfrac{2}{3}$

   D. -2 or $\dfrac{1}{3}$

4. There are three consecutive even integers such that if three-fourths of the smallest is added to the sum of the other two, the result is 39. What are the integers?

    A. 12, 13, 14

    B. 12, 14, 16

    C. 12

    D. 10, 12, 14

5. The difference between a positive number and 14 times its reciprocal is 5. Find the number.

    A. 7

    B. −2

    C. 2

    D. cannot be done with the information given

6. If $1 - \dfrac{3}{x} = 5 + x$, with $x \neq 0$, which of the following is a possible value for x?

    A. 3

    B. −3

    C. 1

    D. −6

7. If $f(x) = x^3 + 5x^2$, the value of $f(-3)$ is $\boxed{}$.

8. If $f(x) = x^2 + 4$, only one of the following numbers can be a value of $f(x)$. Which is it?

    A. 0

    B. −4

    C. 2

    D. 4

9.   The graph of $f(x) = 3x^2 + 5$ could look like which of the following?

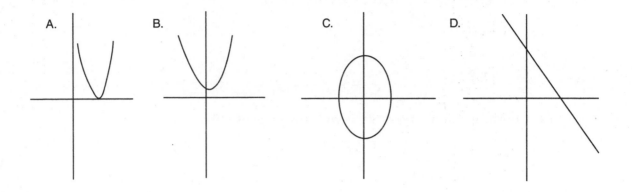

A.   B.   C.   D.

10.   If $f(2) = 5$ and $f(4) = 17$, which of the following can be $f(x)$?

A. $2x + 1$

B. $x + 3$

C. $x^2 + 1$

D. $x^3 - 3$

11.   Simplify: $\dfrac{x^2 - 2x - 15}{x^2 + 4x + 3}$.

A. $\dfrac{x+5}{x+1}$

B. $\dfrac{x-5}{x+1}$

C. $\dfrac{x-5}{x-1}$

D. cannot be simplified

12.   Consider the function $F(x) = x^2 - 3x + 7$. The value of $F(1)$ $\left\{\begin{array}{c}\text{is less than}\\ \text{equals}\\ \text{is more than}\end{array}\right\}$ $F(2)$,

and $F(2)$ $\left\{\begin{array}{c}\text{is less than}\\ \text{equals}\\ \text{is more than}\end{array}\right\}$ $F(3)$.

13. What is the value of $20x - 3y^2$ if $x = 4$ and $y = 2$?

    A. 0

    B. -8

    C. 68

    D. 92

14. In the graph of $f(x) = 2^x$, what is the value of $f(0)$?

    A. 0

    B. 1

    C. 2

    D. -1

15. Jeanette opened a cookie shop in her neighborhood. At first, sales were flat, but then all of a sudden, according to Jeanette, her shop became so popular that she made $100 more every day for the past week. Her daily net income before this spike was a mere $50 per day.

    a. Which of the following equations is the algebraic representation of Jeanette's net income this past week, where x is the day of the week, starting at day 1?

        A. $f(x) = x^2 + 100x + 50$

        B. $f(x) = 100x + 50$

        C. $f(x) = x + 150$

        D. $f(x) = 100x$

    b. Fill in the following table showing Jeanette's net income per day.

Day 1	$150
Day 2	$250
Day 3	$ ☐
Day 4	$450
Day 5	$ ☐

Solutions

Answer 1. $\pm 3\sqrt{2}$. This problem can be done in two ways. If you consider 18 a "square" whose square root is $\sqrt{18}$, then this equation is the difference of squares. This should be considered whenever a quadratic equation has no x term and has a minus sign. Then this can be factored as the difference of two squares: $(x + \sqrt{18})(x - \sqrt{18}) = 0$, and the roots are $x = \pm\sqrt{18} = \pm\sqrt{9 \times 2} = \pm 3\sqrt{2}$. But a quadratic can also always be solved by using the quadratic equation, which gives us

$$x = \frac{-b \pm \sqrt{b^2 - 4ac}}{2a} = \frac{0 \pm \sqrt{18}}{2} = \pm\sqrt{18} = \pm\sqrt{9 \times 2} = \pm 3\sqrt{2} \text{ as well.}$$

Answer 2. (C) ± 10. To solve a quadratic equation, first put all the terms on one side of the equation with 0 on the other side. Then this problem becomes $2x^2 - 200 = 0$, or $x^2 - 100 = 0$ by dividing by 2, which is the difference of two squares; thus, $(x + 10)(x - 10) = 0$, and $x = \pm 10$. If you didn't recognize this as the difference of two squares, you could always use the quadratic formula with $a = 2$ and $c = 200$,

which gives you $x = \dfrac{0 \pm \sqrt{1,600}}{4} = \dfrac{\pm 400}{4} = \pm 10$. This problem can also be solved

by substituting in the answer choices, but watch out! Even though answer choice (B) (10) works, remember that a quadratic needs two answers, so (C) is the correct answer.

Answer 3. (B) 2 or $-\dfrac{1}{3}$. $3x^2 - 5x - 2 = 0$, so $a = 3$, $b = -5$, and $c = -2$.

Using the quadratic equation, $x = \dfrac{5 \pm \sqrt{49}}{6} = \dfrac{5 \pm 7}{6} = 2$ or $-\dfrac{1}{3}$.

Answer choices (A) and (C) indicate that, for the radical, you may have gotten $\sqrt{25 - 24} = 1$ instead of $\sqrt{25 + 24} = 7$. Remember that a minus times a minus is a plus.

Answer 4. (B) 12, 14, 16. This is an example of a problem for which substituting the answer choices into the problem may be the fastest way to go. But first, you should recognize that answer choice (A) cannot be the answer because the integers are supposed to be even, and (C) cannot be the answer because the question asked for all three numbers. When you substitute the numbers in answer choice (B) into the problem, you see that it is the correct answer. This problem also could be solved by writing an equation, but you have to realize that three consecutive, even integers have to be represented by x, $x + 2$, and $x + 4$. Then $\dfrac{3}{4}x + (x + 2) + (x + 4) = 39$ is the equation. Multiply every term on both sides by 4 to get rid of the fraction, and you get $3x + 4(x + 2) + 4(x + 4) = (4)39$, which, after removing the parentheses and combining like terms, becomes $11x = 132$, and $x = 12$. Then the other two numbers are 14 and 16 because x is the smallest number. It's much easier in this case to use elimination of two answer choices and then check the other two.

Answer 5. (A) 7. This problem is a great problem because it contains so many of the tips and concepts that have been presented so far. First, you must know the meaning of the word *reciprocal*. Then you must recognize that the words *the difference between* indicate subtraction, usually the first number mentioned minus the second number mentioned, but watch the wording. And you must recognize that answer choice (B) is wrong because the problem asks for a positive number.

The easiest way to do this (and many other problems involving numbers) is to substitute the answer choices into the information in the problem. But if you were to work it out, the equation is

$$x - 14(\frac{1}{x}) = 5.$$

Multiply through by x $\qquad\qquad$ $x^2 - 14 = 5x.$

Move everything to one side $\qquad$ $x^2 - 5x - 14 = 0.$

Use the quadratic formula $\qquad$ $x = \dfrac{5 \pm \sqrt{81}}{2} = \dfrac{5 \pm 9}{2} = 7$ or $-2.$

Answer 6. (B) -3. Again, with such small numbers, substituting the answer choices into the equation is the fastest way to do this problem. If you choose to do the algebra, the first step is to multiply every term on both sides by x to clear the equation of fractions. Then you get $x - 3 = 5x + x^2$. When all of the terms are put on one side of the equals sign, the equation is $x^2 + 4x + 3 = 0$, and

$$x = \frac{-4 \pm \sqrt{4}}{2} = \frac{-4 \pm 2}{2} = -3 \text{ or } -1.$$

Answer 7. 18. If $f(x) = x^3 + 5x^2, f(-3) = (-3)^3 + 5(-3)^2 = -27 + 45 = 18.$

Answer 8. (D) 4. We can rewrite $f(x) = x^2 + 4$ as $x^2 = f(x) - 4$. Since x^2 must be positive, $f(x)$ must be ≥ 4.

Answer 9. (B). The equation is quadratic and thus the graph should be a parabola, either answer choice (A) or (B). If you substitute $x = 0$ into the equation, $f(x) = 5$, which eliminates choice (A).

Answer 10. (C) $x^2 + 1$. Although all of the answer choices work for $f(2)$, only (C) works also for $f(4)$.

Answer 11. (B) $\dfrac{x-5}{x+1}$. The numerator and denominator can be factored:

$\dfrac{x^2-2x-15}{x^2+4x+3} = \dfrac{(x+3)(x-5)}{(x+3)(x+1)} = \dfrac{x-5}{x+1}$. If you use the quadratic equation to get the factors, remember that it gives you the roots, and then the factors are $(x - \text{root})$. So for

$x^2-2x-15,\ x = \dfrac{-(-2)\pm\sqrt{(-2)^2-4(1)(-15)}}{2} = \dfrac{2\pm\sqrt{64}}{2} = \dfrac{2\pm 8}{2} = 5 \text{ or } -3$,

so the factors are $(x+3)$ or $(x-5)$. And for x^2+4x+3,

$x = \dfrac{-(4)+\sqrt{(4)^2-4(1)(3)}}{2} = \dfrac{-4\pm\sqrt{4}}{2} = \dfrac{-4\pm 2}{2} = -3 \text{ or } -1$, so the factors are

$(x+3)$ or $(x+1)$.

Answer 12. equals; is less than. $F(1) = 1^2 - 3(1) + 7 = 5$. $F(2) = 2^2 - 3(2) + 7 = 5$. $F(3) = 3^2 - 3(3) + 7 = 7$.

Answer 13. (C) 68. The value is $20(4) - 3(2)^2 = 80 - 12 = 68$.

Answer 14. (B) 1. The value of $f(0) = 2^0$, and anything to the zero power is 1.

Answer 15. a. (B) $f(x) = 100x + 50$. Answer choice (A) is a quadratic, so it should be eliminated right away. Answer choice (C) would give a value of $151 for day 1, so it should be eliminated. Answer choice (D) doesn't include the fact that Jeanette is talking about her increase in income being above her expected $50 per day of the week before.

b. $350; $550. Jeanette is making $100 more each day.

The Shape of Things

This chapter is about geometry. Every day, you see many things that have to do with geometry and you use geometric principles, even though you don't think of them as geometry. Tires are circles, and they had better be attached at the exact center of the circle to function properly. Honeycombs are made up of hexagons (six-sided figures). Even the truss on a bridge is a trapezoid, and bridges are made up of many triangles because the triangles create rigidity.

A lot of understanding geometry is knowing the words that describe a shape. Pay particular attention to the definitions in the following sections, although they are words you probably already know.

Two words that pertain to all two-dimensional closed geometric figures are perimeter and area. (**Closed** means all the corners are connected.) The **perimeter** is the distance around a figure, or the sum of the lengths of all of its sides. A typical perimeter is a fence around a plot of land. **Area** is a term used for the space enclosed by any closed figure. It is expressed in square units (in^2, ft^2, and so forth) and is found by various formulas, some of which are on the GED® test formula sheet. Typical areas that we see every day are a rug or a plot of land enclosed by a fence.

Lines and Angles

Geometric shapes have everything to do with lines and angles, so you must understand them first. Even circles, which themselves have no straight lines or angles, have straight lines and angles within them that tell, for example, the size of the circle as well as parts of the circle.

A **line** actually goes on forever in both directions, or we say, "It goes on to infinity (∞) in both directions." If we want to concentrate on a part of a line, we call that a **line segment**, and we show which line segment we mean by stating its **endpoints**. So if we are interested in a line that goes from the 1-inch to the 5-inch measure, we mean a 4-inch line segment. A line segment, since it has a definite measure, has a **midpoint**, which is exactly halfway between the endpoints. A **ray** goes off to ∞ in only one direction. The other end has a definite point, or endpoint.

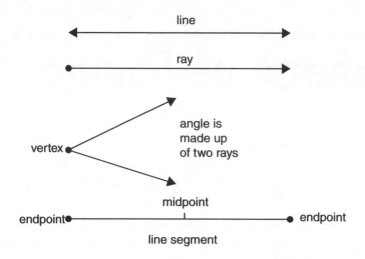

If we put two line segments or rays together at one of their endpoints, at a point called the **vertex**, we get an **angle**. The measure of the angle tells how wide open it is. If the angle is all the way open and forms a straight line, it is called a **straight angle**, and its measure is 180°. Three other types of angles, named for their measurements, are all less than 180°.

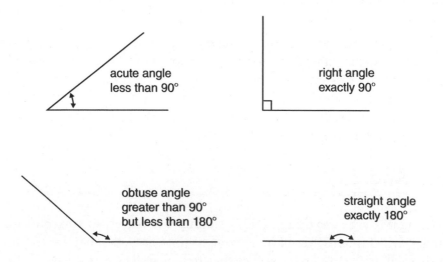

As you can see, if we have several angles with the same vertex that form a straight line, their measures must add up to 180°. But if one of the adjacent angles is obtuse, it is the only obtuse angle because it uses up more than 90° of the 180° in a straight line.

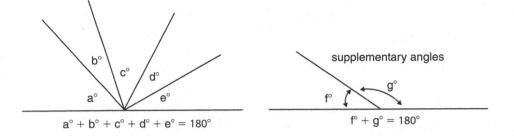

$a° + b° + c° + d° + e° = 180°$

supplementary angles

$f° + g° = 180°$

Two angles whose measures add up to 180° are called **supplementary** angles. They don't have to be adjacent angles, like the ones shown in the figure above (right), they just have to add up to 180°. Similarly, two angles whose measures add up to 90° are called **complementary** angles, and again they don't have to be adjacent. The next figure shows adjacent and nonadjacent pairs of complementary and supplementary angles.

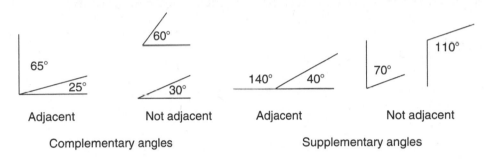

Can you see that if two adjacent angles are equal and supplementary, they must be two right (90°) angles (and the lines are therefore perpendicular)? What would be the measures of two adjacent angles that are equal and *complementary?* The answer is 45° each because 45° + 45° = 90°.

Now let's consider other angle relationships. If two lines **intersect** (cross each other), they form two pairs of **vertical angles**, which is the name for the angles across from each other. Pairs of vertical angles are equal. Suppose in the following figure that we know $\angle A$ is 30°. Then $\angle C$ is also 30° because it is a vertical angle. Can we also figure out the measures of $\angle B$ and $\angle D$? Yes, because $\angle A$ and $\angle B$ together form a straight line, so they must add up to 180°. So $\angle B$ must be 150° (that is 180° − 30°), and then its vertical angle, $\angle D$, has to be 150° also. So the angles in the figure below are either of only two angle measurements, 30° and 150°.

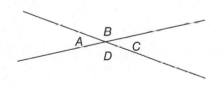

Two parallel lines that are intersected by a third line (called a transversal) form eight angles that have special relationships based on vertical angles and supplementary angles. Let's add a parallel line (line 2) to the above figure (where the original line is line 1):

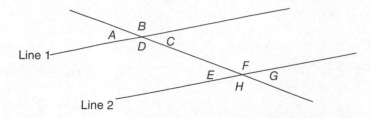

We already know that $\angle A = \angle C$, and $\angle B = \angle D$, but what about the other group of angles? They are also vertical and supplementary angles, so similarly, $\angle E = \angle G$, and $\angle F = \angle H$. Later, when we talk about parallelograms, we will see, in addition, that $\angle A = \angle E$. For now, we will just say this is true, and if it is true, then $\angle A = \angle C = \angle E = \angle G$. Similarly, then $\angle B = \angle D = \angle F = \angle H$. So in this entire figure, there are only two measures for all eight angles. If we say $\angle A$ is 30°, the other measure is 150°. There are other names for the relationships among these angles besides being called vertical and supplemental angles, but we only have to know that if a third line crosses two parallel lines, the angles that are formed have one of only two measures.

Triangles

Triangles are formed when three line segments form a closed figure. Triangles are one of the most important shapes in geometry. One of the properties of triangles that you probably didn't think about until now is that they are rigid. That means that once you have three sides attached in a triangle shape, you cannot change it. You cannot make one of the sides longer, and you cannot make any of the angles larger without changing all sides. This is why bridge supports have triangles as their major structure.

Try this for yourself: Take two soda straws. Bend one into a four-sided figure and the other into a triangle. Hold together the angle where the ends of each straw meet. Now apply pressure on one of the other sides or angles. The four-sided figure collapses on itself, but the triangle stays rigid.

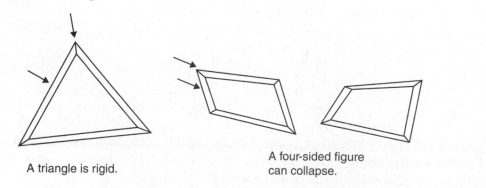

A triangle is rigid.

A four-sided figure can collapse.

There are several kinds of triangles, but they all have the following things in common:

1. They have three sides and three angles. In fact, that is what the word *triangle* means: three angles.

2. The largest angle is across from the longest side and the smallest angle is across from the shortest side.

3. The measures of the three angles add up to 180°. So if you know the measures of two angles of a triangle, you can always find the third. This is a consequence of the rigidity of a triangle.

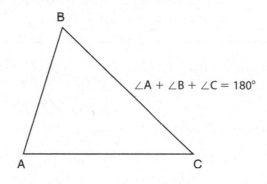

4. Another fact that has to do with a triangle's rigidity is that, if you know two sides and the angle between them, the other side and the other two angles are fixed. Finding the measures requires a knowledge of trigonometry, which is beyond what the GED® test covers. Can you see that once we have the lengths of sides *DE* and *DF* and we know ∠*D*, there is only one way to form a triangle (draw *EF*)?

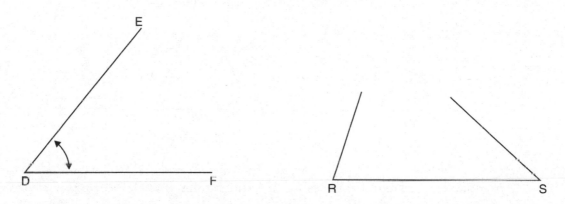

5. If you know one side and two angles in a triangle, the lengths of the other two sides are predetermined. Can you see above (right) that if we extend the two partial sides, they will meet at a point (let's call it *T*) that depends on ∠*R* and ∠*S*? Thus, there is one and only one triangle that can be drawn if you know the two angles and the distance between them. Again, finding the lengths of any of the two remaining sides requires a knowledge of trigonometry.

Note that if all you know are the lengths of two sides of a triangle and nothing about the angles, you don't automatically know the third side of the triangle—it actually could be anything, depending on what the angle between the two sides is.

HINT

In the figures in this chapter, if sides are equal, they will be marked with the same tick marks (either one or two). Likewise, if angles are equal, they will be marked with the same angle mark. Right angles are marked with a small box.

Types of Triangles

Triangles can be classified by their side measurements.

Scalene: A triangle in which the lengths of all of the sides are different.

Isosceles: A triangle with two equal side lengths. The two equal sides are called the **legs**, and the third side is called the **base**.

Equilateral: All three side lengths are equal. The name comes from the two parts of the word: *equi* ("equal") and *lateral* ("sides").

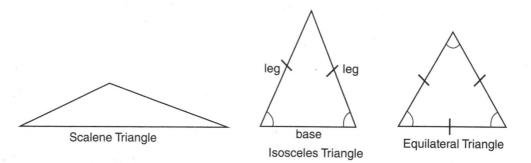

Scalene Triangle

Isosceles Triangle

Equilateral Triangle

An equilateral triangle is also called an **equiangular** triangle because, if all three sides are equal, so are all three angles. The two parts of the word *equiangular* are: *equi* ("equal") and *angular* ("angle"). The angles in an equilateral triangle are each 60° because there are three of them and they have to add up to 180°.

The sum of the lengths of any two sides of a triangle must be greater than the length of the longest side, or we don't have a triangle (the two sides won't meet). Another way to look at this rule is that the difference between the lengths of any two sides of a triangle must be less than the length of the shortest side.

A way to remember that the sum of the lengths of any two sides of a triangle must be greater than the length of the longest side is by visualizing a triangle with sides 1, 2, and 4 (see below). Can it form a triangle? No, the sides that are lengths 1 and 2 would just fall down on the side that is 4 and never meet at the third angle. Their sum has to be more than 4 (the longest side). Likewise, the side of length 1 (shortest side) is too short to form a triangle with the sides of length 2 and 4. If it were longer than 2 (the difference of 4 − 2), then it could form a triangle.

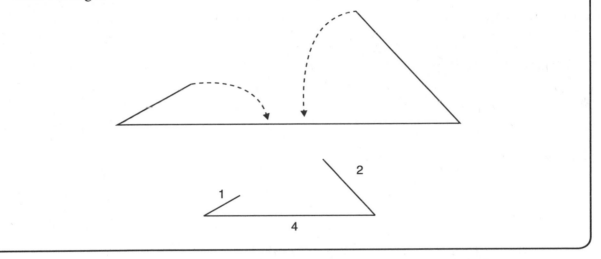

Triangles also can be classified by their angle measurements.

Acute: All the angles are acute angles.

Obtuse: One of the angles is an obtuse angle. There can be only one obtuse angle in a triangle because it uses up more than 90° of the total 180° in a triangle.

Right: One of the angles is a right angle. This is such an important kind of angle that it deserves a section in this chapter all its own.

Example 7.1.

You are given $\triangle ABC$ with sides $AB = 4$ and $BC = 6$.

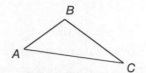

The length of side AC cannot be which of the following numbers?

    A. 8

    B. 10

    C. 5

    D. 3

Answer 7.1.

(B) 10. The third side must be less than 10 (the sum $AB + BC$) but greater than 2 (the difference $AB - BC$).

Example 7.2.

If the measures of two of the angles of an isosceles triangle are 30° and 120°, which of the following must be the measure of the third angle?

    A. 120°

    B. 90°

    C. 60°

    D. 30°

Answer 7.2.

(D) 30°. Since it is an isosceles triangle, this angle must be the same as one of the given angles. But since a triangle can have only one obtuse angle, it must be 30°. Another way to look at this problem is that the angles of a triangle always total 180°, so $180 = 120 + 30 + x$, and $x = 30°$.

Example 7.3.

In this figure (not drawn to scale), what is the measure of ∠A?

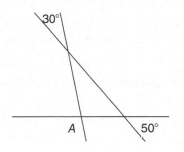

A. 100°

B. 80°

C. 20°

D. cannot be determined with the information given

Answer 7.3.

(A) 100°. Because vertical angles are equal, two of the angles of the triangle are 30° and 50°, which add up to 80°, leaving 100° for ∠A because the sum of angles of a triangle is 180°. And since the third angle is 100°, so is its vertical angle, ∠A.

Right Triangles

Right triangles are a special class of triangles because they are so common in our everyday life. They get their name from the fact that one of the angles is a right angle, or 90°, which means that two of the sides are perpendicular. Everyday examples of right triangles include a ladder propped against a house (the house is perpendicular to the ground) or a shadow cast by a tree (the tree is perpendicular to the ground).

In a right triangle, you need to know only one of the other angles to figure out the third angle because you know the right angle is 90° and the measures of the three angles in a triangle add up to 180°. Thus, the other two angles add up to 90°. For example, in a right triangle with one angle of 30°, the third angle is found by subtracting the other angle from 90°:

$$90° - 30° = 60°.$$

Remember, though, that this quick subtraction from 90° applies only to *right* triangles.

The sides of a right triangle are special, too. Knowing two sides and automatically being able to find the third side isn't possible with any other triangle unless you also know one of the angles. But in a right triangle, you automatically know one of the angles—it is 90°. In fact there is a formula for finding the third side of a right triangle if you know the other two sides.

That formula is from the **Pythagorean Theorem**, which states, "The square of the hypotenuse is equal to the sum of the squares of the other two sides." Some definitions are needed here before we write the formula (which is on the formula sheet of the GED® test). The **hypotenuse** of a right triangle is the side opposite the 90° angle. It is the longest side. The **legs** are the other two sides. The formula thus is written as

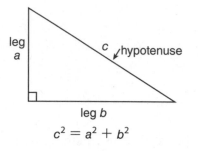

$$c^2 = a^2 + b^2$$

The Pythagorean Theorem can be used to find the distance between two points on the Cartesian coordinate plane (see Chapter 4). The two legs are found by calculating the differences in the x values ($x_2 - x_1$) and the y values ($y_2 - y_1$). Then use these values as the a and b values, respectively, in the Pythagorean Theorem, and the c value will give the distance between the points. This can be seen in the following figure.

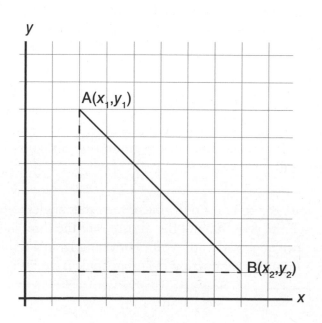

Even though the Pythagorean Theorem appears on the formula sheet of the GED® test, it is a good idea (and a time-saver on the test) if you know at least one of the **Pythagorean triples**. These are the sides of right triangles that are whole numbers, and thus easy to work with. The most popular is easy to remember. It is known as the 3-4-5 right triangle. You can see that $5^2 = 3^2 + 4^2$, so it is indeed a right triangle.

HINT

Any multiple of the 3-4-5 right triangle is also a right triangle (such as 6-8-10 or 9-12-15). If you know the two legs of a right triangle are 6 and 8, you automatically know the third side is 10. Other Pythagorean triples are not as easy to remember as the 3-4-5. They are 5-12-13 and 8-15-17 and, of course, any multiples of them.

Example 7.4.

A ladder 10 feet long is propped 8 feet up on a house. The base of the ladder is ⬚ feet from the foundation of the house.

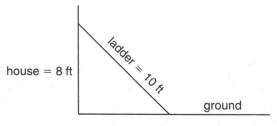

house = 8 ft   ladder = 10 ft   ground

Answer 7.4.

6. Since this is a right triangle (the house and its foundation form a right angle), we can use the Pythagorean Theorem: $c^2 = a^2 + b^2$, or $10^2 = 8^2 + b^2$, so $b^2 = 36$, and $b = 6$. If you recognized that this is double a 3-4-5 right triangle, you know its sides are 6-8-10. Since the problem already has "feet" after the answer, write in only the 6.

Example 7.5.

To drive to the ball game after work, Nick drives 3 miles north and then 4 miles west. The ball field is ⬚ miles from Nick's workplace.

5. This forms a right triangle in which the distance from the ballfield to work is the hypotenuse. Use the Pythagorean Theorem: $c^2 = 3^2 + 4^2 = 25$. So the ball field is 5 miles from Nick's workplace. If you recognized this as a 3-4-5 right triangle, you got the answer right away. Put only 5 in the blank because "miles" is already in the sentence.

Perimeter and Area of a Triangle

The **perimeter of a triangle** is simply the sum of the side lengths. If the triangle is an isosceles triangle, two lengths will be the same. If it is an equilateral triangle, the perimeter is three times the length of one side because they are all the same. For a right triangle, you may have to calculate the third side by the Pythagorean formula before you can add the three sides. So for any triangle with three sides, a, b, and c, the perimeter (p) is given by

$$p = a + b + c.$$

The **area of a triangle** is one-half the base times the height of the triangle. You can choose any of the three sides to be the **base**, although for an isosceles triangle it is easiest if it is the unequal side, and for a right triangle it should be one of the legs. The **height** is the *perpendicular* distance from the base to the opposite angle.

HINT

This is the tricky part: finding the height of a triangle. It is one of the sides only if the triangle is a right triangle because the two legs are perpendicular. So for any other triangle, remember that it is not one of the other sides. In fact, for an obtuse triangle, the height could be a measurement outside the triangle itself.

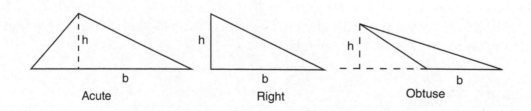

Acute   Right   Obtuse

The formula for the area (A) of a triangle, where b is the base and h is the height to that base, is

$$A = \frac{1}{2}bh.$$

Where did the $\frac{1}{2}$ come in? For any triangle, we can duplicate it across any side and we will end up with a four-sided figure *in which the sides across from each other are equal*. We will see shortly that the area of this new four-sided figure (called a parallelogram) is simply its base times its height (or length times width). The two triangles are identical. The area of either triangle is one-half the area of the four-sided figure. So when you figure the area of a triangle, don't forget the $\frac{1}{2}$.

HINT

Even though a formula sheet is provided on the GED® test, the formulas for the perimeter and area of a triangle (and a few others) are not listed there but you are expected to know them. At the end of this chapter is a list of the formulas you are expected to know and that don't appear on the GED® test formula sheet but may appear on the GED® test.

Because all equilateral triangles always have 60° angles, they are exactly the same shape. The only thing that is different from one equilateral triangle to another is its size. Imagine two equilateral triangles, one with a side of 1 inch and the other with a side of 2 inches. The **scale factor** between these two triangles is two. The perimeter of the larger triangle (6 inches) is two times the perimeter of the smaller triangle (3 inches). What about the areas of the two triangles? Since area involves the product of two dimensions ($\frac{1}{2} \times$ base $\times$ height), and each of them in the smaller triangle is scaled by two in the larger triangle, the area is enlarged 2×2, or 4 times.

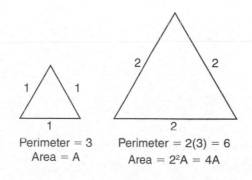

Perimeter = 3
Area = A

Perimeter = 2(3) = 6
Area = 2²A = 4A

HINT

If two two-dimensional figures have the same shape and are scaled by a factor of x, their areas are scaled by a factor of x^2. The fact that area measures are given in square units will remind you to square the scale factor.

Example 7.6.

The longest side of a right triangle is 13 inches and the shortest side is 5 inches.

   a. The perimeter of the triangle is

       A. 12 inches

       B. 30 inches

       C. 7 inches

       D. 20 inches

   b. The area of the triangle is

       A. $32\frac{1}{2}$ square inches

       B. 78 square inches

       C. 30 inches

       D. 30 square inches

Answer 7.6

   a. (B) 30 inches. This is a right triangle, so the longest side is the hypotenuse. Use the Pythagorean Theorem to find the third side: $13^2 = 5^2 + b^2$. Then $b^2 = 144$, and the third side is 12 inches. The perimeter is then the sum of the sides, or $13 + 5 + 12 = 30$ inches.

b. (D) 30 square inches. The legs of the right triangle are 5 and 12, as determined in part (a). Then the area is $\frac{1}{2}(5)(12) = 30$ square inches. Area is measured in square units, so answer choice (C) is incorrect.

Quadrilaterals

Quadrilaterals (*quadri* = "four" and *lateral* = "side") are four-sided closed figures. We are familiar with squares and rectangles, but there are infinitely many other quadrilaterals. Their only requirement is that they be closed and have four sides.

Two facts are true of all quadrilaterals:

1. The sum of the angles is 360°.

2. The perimeter of a quadrilateral is the sum of the lengths of the four sides.

Five quadrilaterals are special due to facts about their angles and side lengths, as shown in the next five subsections.

Trapezoid

A **trapezoid** is a quadrilateral with only one special fact: two sides are parallel. They don't have to be equal, and the other two sides don't have to be parallel or equal either.

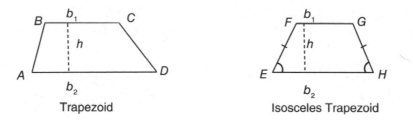

However, if the other two sides are equal, just as for triangles, the trapezoid has a special name: isosceles. An **isosceles trapezoid** has two parallel sides (*FG* and *EH* in the figure above), and the other two sides (*EF* = *HG*) are equal. The base is usually chosen to be the longest parallel side, and the base angles (the angles on either side of it) are equal ($\angle E = \angle H$), although it is also true that the other two angles are equal ($\angle F = \angle G$). In addition, for an isosceles trapezoid, the diagonals (here they would be *EG* and *FH*) are of equal length.

The *perimeter* of any trapezoid is just the sum of the sides ($AB + BC + CD + DA$); for the isosceles trapezoid, it is ($EF + FG + GH + HE$). So for a trapezoid with four sides of lengths, say, a, b, c, and d, the perimeter is

$$p_{\text{trapezoid}} = a + b + c + d$$

The area of any quadrilateral is based on the simple formula of base $\times$ height (also called altitude), but for a trapezoid we have to consider the average of the two bases as the base in this calculation. Otherwise, the area would be two different numbers, depending on which of the parallel sides is considered to be the base. The height is defined as perpendicular to the base, the same as for triangles. Since the bases are parallel to each other, the height is perpendicular to each base and its size doesn't vary. How do we determine the average of the two bases? Averages are covered in Chapter 8, but basically the average of two quantities is their sum divided by 2. So the formula to use for a trapezoid is $\frac{1}{2}(b_1 + b_2)$, where b_1 and b_2 are the lengths of the two parallel sides (bases). The area of a trapezoid is given by

$$A_{\text{trapezoid}} = \frac{1}{2}(b_1 + b_2)h.$$

This formula is the same as the one provided on the GED® test formula sheet: $A = \frac{1}{2}h(b_1 + b_2)$ since multiplication is commutative (xy is the same as yx; see Chapter 2).

Parallelogram

A **parallelogram** is a quadrilateral (four-sided figure) with parallel sides, as the name implies. In this case, it is a step up from the trapezoid because the other two sides are also parallel. So a parallelogram has two pairs of parallel sides, and the parallel sides are equal to each other.

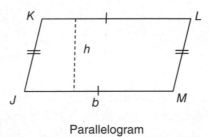

Parallelogram

The *perimeter* of a parallelogram is still the sum of the lengths of all four sides. So the perimeter of the above parallelogram is $JK + KL + LM + MJ$. In general, if the lengths of the sides of a parallelogram are, say, q, r, s, and t, its perimeter is

$$p_{\text{parallelogram}} = q + r + s + t.$$

Since the sides opposite each other are equal as well as being parallel, if $q = r$ and $s = t$, then the perimeter can be rewritten as

$$P_{\text{parallelogram}} = 2q + 2s = 2(q + s).$$

Thus, the perimeter of a parallelogram is twice the sum of the two unequal sides.

The *area* of a parallelogram is given by the general formula of base × height, where the height is perpendicular to the base:

$$A_{\text{parallelogram}} = bh$$

The diagonals of a parallelogram bisect each other. A new feature of the parallelogram is that the angles form two pairs, with the ones across from one another being equal. The pairs of adjacent angles ($\angle J$ and $\angle K$, $\angle K$ and $\angle L$, $\angle L$ and $\angle M$, or $\angle M$ and $\angle J$ in the figure) add up to 180°, so they are **supplementary** angles.

Rhombus

If we add the condition that all four sides of a parallelogram are equal, then we have a **rhombus**. So a rhombus has all of the properties of a parallelogram plus the sides are equal.

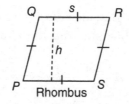
Rhombus

Thus, in the above figure of the rhombus, $PQ = QR = RS = SP = s$, and the *perimeter* can be written as

$$P_{\text{rhombus}} = 4s$$

Likewise, the *area* of the rhombus is

$$A_{\text{rhombus}} = bh = sh,$$

where any of the sides can be used as the base, and the height drawn to each side is the same.

The diagonals of a rhombus bisect each other (as they did for the parallelogram), but now they also are perpendicular to each other.

Rectangle

If, instead of saying the four sides of the parallelogram are equal, we say that the four angles are equal, we have a **rectangle**, which is a parallelogram with four equal angles.

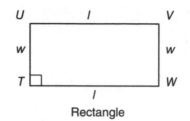

Rectangle

Thus, in the figure of the rectangle above, $\angle T = \angle U = \angle V = \angle W$, and since the angles of a quadrilateral add up to 360°, each of the four angles is 90°, or a right angle.

The opposite sides are equal, as in a parallelogram, but not all sides are equal (as they were in the rhombus). The *perimeter* is written as usual as

$$P_{rectangle} = TU + UV + VW + WT = 2l + 2w = 2(l + w).$$

Because all the angles are right angles, all sides l (length) are perpendicular to sides w (width), so they take the place of the base and height, and the *area* of the rectangle is

$$A_{rectangle} = bh = lw.$$

$\mathsf{H}\mathsf{INT}$

If we think of tiling a floor in a straight line, we get an idea of why area is length times width. Suppose we wanted to tile a room that is 12 feet by 10 feet in 1-foot tiles. If you count the number of tiles needed, it would be 12 rows of 10 tiles (or 10 rows of 12 tiles), and that total is 120 tiles, each 1 foot square, which is the area of the floor: 120 square feet.

The diagonals of a rectangle are equal, and they bisect each other (as they did for the parallelogram), but they are not perpendicular to each other—that was true only for the rhombus.

We can halve a rectangle in several ways, and the area of each half will always be half the area of the original rectangle, even though they may have different shapes.

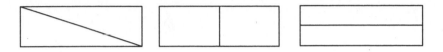

Square

Now we come to the most restrictive, but also the most popular, quadrilateral—the square. A square is a parallelogram with four equal angles and four equal sides. It can also be thought of as a rhombus with four equal angles or a rectangle with four equal sides.

Therefore, the square has all of the properties of the parallelograms mentioned above, as shown in the following table.

Properties of the Square

Property of the square:	Same as property for:
Opposite sides are parallel	Parallelogram
All sides are equal	Rhombus
All angles are equal	Rectangle
Diagonals bisect each other	Parallelogram
Diagonals are perpendicular to each other	Rhombus
Diagonals have equal lengths	Rectangle

Since a square is a rectangle with equal sides, its perimeter is

$$P_{square} = 4s,$$

and its area is

$$A_{square} = s^2.$$

Example 7.7.

The value of *x* in the quadrilateral below is

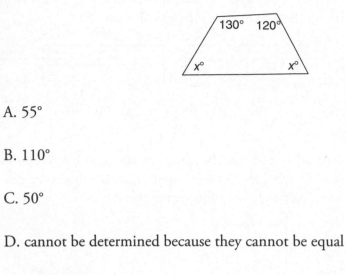

A. 55°

B. 110°

C. 50°

D. cannot be determined because they cannot be equal

Answer 7.7.

(A) 55°. All of the angles in a quadrilateral total 360°. So we have

$$130° + 120° + x + x = 360°$$

$$2x = 110°$$

$$x = 55°$$

The problem is to find the value of *x*, not 2*x*. Also, unless you are told what a figure is, don't assume it is what looks like it. This figure looks like a trapezoid, but it isn't. Answer choice (D) is true for a trapezoid, but not this figure, which is just a quadrilateral, as the problem states.

Example 7.8.

Which of the following rectangles have the same perimeter? (The figures are not drawn to scale; use the given measurements.)

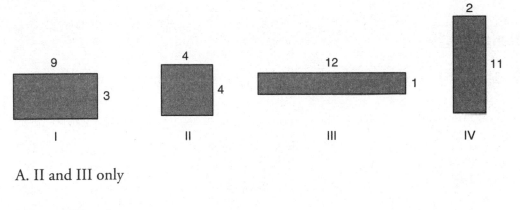

A. II and III only

B. I and III only

C. III and IV only

D. all of them

Answer 7.8.

(C) III and IV only. The perimeter of a rectangle is $p = 2(l + w)$. (This formula is not on the GED® test formula sheet.) For figure I, $p = 24$; for II, $p = 16$; for III, $p = 26$; and for IV, $p = 26$. Therefore, the correct answer choice is (C).

Example 7.9.

Mr. Hermann wants to store some books in an office that is 12 feet by 15 feet. If one entire wall is to be used for bookshelves that are 1 foot deep, on which wall should he place them so the remaining floor area is a maximum?

A. On the shorter wall

B. On the longer wall

C. The remaining floor space will be the same for either wall.

D. Cannot determine with the information given

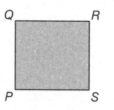

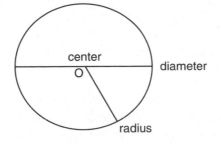

Another important value when talking about circles is the Greek letter pi (π), which is the ratio of the circumference (perimeter) of the circle to its diameter. Pi has a value of $\dfrac{22}{7}$, or roughly 3.14. The GED® test will indicate what form to use for an answer, but it usually is in terms of π.

CALCULATOR

The calculator has a button labeled π, and you can treat π just as you would any constant. That is, to total $2\pi + 3\pi$, you can press those buttons, with "enter" and the value 5π appears. If you then use the toggle button, the calculator will give the decimal equivalent, but that usually isn't required.

The angle with the center of a circle at its vertex and two radii (plural of *radius*) as its sides is called a **central angle**. The sum of the central angles in a complete rotation (or circle) is 360°. Think of the hands of a clock. At 3:00, the angle between the hands on a clock is 90° (one-fourth of the way around the whole clock face). It is also 90° at 9:00.

The portion of the circumference of a circle between two points is called an **arc**. Its measurement is the same as the measure of the central angle drawn to it (see the figure below).

Be careful when figuring the angle between the hands of a clock. A clock shows 12 hours, but the angle between any two numbers on a clock is 360° ÷ 12 = 30°, at 4:10, the angle between the hands of the clock is not 60° (the angle between the 2 and the 4) because the hour hand will have moved beyond the 4, on its way to the 5.

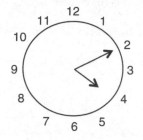

The *perimeter* of a circle is called its **circumference**, and the formula for it is

$$C = \pi d,$$

where *d* is the diameter. Since the diameter is twice the radius (*r*), *d* = 2*r*, the circumference can also be written as $C = 2\pi r$.

The *area* of a circle is given by

$$A = \pi r^2.$$

"Cherry pie is delicious; apple pies are too."

What does this have to do with circles?

Cherry pie is delicious; → $C = \pi d,$

apple pies are too → $A = \pi r^2.$

This sentence is just a quick way to remember these two equations, which you are expected to know, but they are not included on the GED® test formula sheet.

A **semicircle** is a half circle. Its area is exactly one-half of the area of the full circle. The perimeter, however, is not one half of the circumference of the whole circle, which is seen in the following figure of a semicircle. On this figure, the perimeter is the round part (which is half of the circumference of the whole circle), plus the diameter, which is the straight part of the semicircle. Semicircles come up when considering composite figures, the topic of the next section.

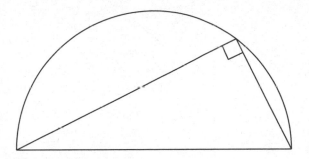

If a triangle is drawn inside a semicircle (or inside a circle with the longest side of the triangle being the diameter), it is a right triangle.

Example 7.11.

Given a circle with a diameter of $\dfrac{2}{3\pi}$ units. Its circumference is $\boxed{}$

units, and its area is $\boxed{}$ square units.

Answer 7.11.

$\dfrac{2}{3}$; $\dfrac{1}{9\pi}$. For the circumference, $C = \pi d$, so $C = \pi\left(\dfrac{2}{3\pi}\right) = \left(\dfrac{\pi}{1}\right)\left(\dfrac{2}{3\pi}\right)$. Cancel the π in the

numerator and denominator to get $C = \dfrac{2}{3}$ units. For the area, we have to find the value of r. Since

$d = 2r$, $r = \dfrac{1}{2}d = \dfrac{1}{2} \times \dfrac{2}{3\pi} = \dfrac{1}{3\pi}$. So the area is $A = \pi r^2 = \pi\left(\dfrac{1}{3\pi}\right)^2 = \dfrac{1}{9\pi}$.

Example 7.12.

The three circles pictured have radii of 3, 4, and 5 inches. What is the perimeter of the triangle formed by connecting the centers of the circles?

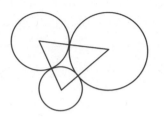

    A. 12 inches

    B. 24 inches

    C. 34 inches

    D. Not enough information is given.

Answer 7.12.

(B) 24 inches. The sides of the triangle are made up of the sum of the radii of the circles. The sides are therefore $3 + 4$, $3 + 5$, and $4 + 5$, or 7, 8, and 9 inches. Then the perimeter is $7 + 8 + 9 = 24$ inches.

Example 7.13.

A circle is drawn inside a larger circle, as shown below. The smaller circle has a diameter of 4 and the larger circle has a diameter of 6. The area of the shaded portion of the figure is [   ] π.

Answer 7.13.

5. The shaded portion is the area of the larger circle minus the area of the smaller circle. Remember to use the radius, not the diameter, to find the area. For the smaller circle ($d = 4$), the radius is 2; for the larger circle ($d = 6$), the radius is 3. Therefore, the shaded portion is $\pi(3)^2 - \pi(2)^2 = 9\pi - 4\pi = 5\pi$. Be sure to enter only the 5 in the answer box since the π symbol is already there.

Composite Two-Dimensional Figures

Composite figures are figures made up of two or more geometric shapes. The perimeter of the shape may be found by adding all of the segments (if there is a circular shape, include the relevant portion of the circumference). The area of the composite shape is found by adding the areas of the composite parts (with no overlapping).

As an example, consider the Norman window shown below, which has the shape of a semicircle on top of a rectangle.

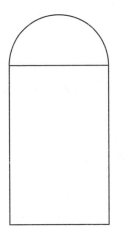

Its perimeter is found by adding the length around the semicircle (curved part only) and three sides of the rectangle. The length around the semicircle is half the circumference of a circle with the same radius. The flat part of the semicircle and the fourth side of the rectangle (the same line) are not part of the perimeter since they are inside the figure. The area of the figure is just the sum of the area of the semicircle and the area of the rectangle.

A topic related to composite figures involves breaking a shape up into triangles or other figures to compute the area (or sometimes the perimeter). Let's look at a parallelogram to see how breaking it up into other shapes can help us to compute the area.

The parallelogram on the left can be broken up into a smaller (darker) rectangle with two triangles on either side (see middle figure). If we then move the triangle on the right over to the left, we end up with the light gray rectangle in the figure on the right, which has the same area as the original parallelogram. Now it is clear why the area of a parallelogram is the same as that of a rectangle with the same base and same height.

Example 7.14.

Consider a Norman window in which the height of the semicircle is 2 feet, and the height of the rectangle is 6 feet.

    a. The perimeter of this window is ☐ feet.

    b. The area of this window is ☐ square feet.

Answer 7.14.

a. $2\pi + 16$. If the height of the semicircle is 2 feet, that is the radius of the semicircle. So the diameter is 4 feet, which is also the width of the rectangle. The perimeter for the round part of the figure is $(\frac{1}{2})4\pi = 2\pi$ because $d = 4$, $C = \pi d = 4\pi$ for the circle, and half that for the semicircle. The dimensions of the rectangle are 4 by 6. The perimeter of the rectangular part of the figure is $6 + 4 + 6 = 16$. Therefore, the perimeter of the Norman window is $2\pi + 16$ feet. (Put only the number in the box; the word *feet* is already included.)

b. $2\pi + 24$. For the area, the semicircle has radius 2 and is half a circle with area $= \pi(2)^2 = 4\pi$, so the area of the semicircle is 2π. The area of the rectangular part is $4 \times 6 = 24$. So the area of the Norman window is $2\pi + 24$ square feet. (Put only the number in the box; the words *square feet* are already included.)

Three-Dimensional Figures

Three-dimensional figures have depth in addition to the length and width that we saw in two-dimensional figures. It is sometimes difficult to visualize three-dimensional figures on a two-dimensional page or computer screen. They are drawn with a certain perspective with which you should become familiar. Solid lines indicate edges that you actually see, and dashed lines indicate edges that are obscured from the two-dimensional perspective. **Edges** are the lines where two **faces** (flat surfaces) meet on a three-dimensional figure. On the figure shown below, the "main" face is shaded and the arrows indicate which way the figure is facing. It takes a while to get used to this way of looking at three-dimensional figures.

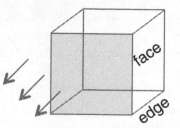

The GED® test formula sheet provides the formulas you will need to do surface area and volume problems with three-dimensional figures. Still, you need to know what these terms mean as well as what the variables in the given equation mean or the formulas won't be of any use to you. Therefore, this section on three-dimensional figures provides information to help you understand the GED® test formula sheet, but it isn't necessary to memorize the formulas.

The **surface area** of a three-dimensional figure is exactly what it sounds like. It is the total area of all the faces, even those you cannot see in the picture. So for surface area, we need to remember the formulas for the areas of the two-dimensional faces that make up each three-dimensional figure. Again, these are given on the GED® test formula sheet. The surface area of a rectangular solid can be thought of as the area of wrapping paper that covers a shirt box with no overlapping.

Volume is how much the three-dimensional figure can hold. It is sometimes called **capacity**. Basically, for three-dimensional figures that have identical "tops" and "bottoms" (bases), it is the area of the **base** (bottom or top) multiplied by the height of the figure.

HINT

Note the dimensions for each measure. Although feet are shown here, you can substitute "inches," "centimeters," or whatever the problem is using.

- Unit   feet   ft

- Surface area   square feet ft^2

- Volume   cubic feet ft^3

Note that the power (exponent) matches the number of the dimensions of the figure.

Rectangular Prisms

One type of three-dimensional figure is known as a **rectangular prism**, which is also referred to as a **right prism**. These figures have perpendicular **edges** and **faces**, and each set of three perpendicular edges meet at a point called the **vertex** (the plural of *vertex* is *vertices*).

Also obscured, but something we can envision, are the diagonals of a rectangular prism, which are not the diagonals of the faces. Rather they are diagonals inside the figure that go from one corner to the opposite corner (for example, from the top left front corner to the bottom right back corner).

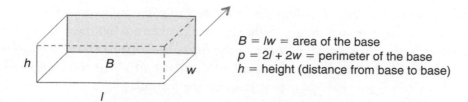

$B = lw$ = area of the base
$p = 2l + 2w$ = perimeter of the base
h = height (distance from base to base)

The *surface area* of a rectangular prism is the sum of the areas of the faces. A rectangular prism has three pairs of identical faces, so we can determine the sum of the areas of the three different faces and double it. Thus, $SA_{\text{rectangular prism}} = 2(lw + lh + wh) = 2lw + 2lh + 2wh$. Another way to find this surface area is to think of the first term in the parentheses (lw) as the area of the base, B, so $2lw = 2B$. Then, if we recognize that $2l + 2w$ is the perimeter, p, of the base, the last two terms can be written as ph. Thus, the surface area becomes

$$SA_{\text{rectangular prism}} = ph + 2B,$$

which is the formula given on the GED® test formula sheet. Just remember what the B, p, and h stand for.

The *volume* of a rectangular prism follows the rule that volume is the area of the base times the height. As we just saw, the area of the base is lw. The volume is thus

$$V_{\text{rectangular prism}} = lwh = Bh,$$

which is the formula given on the GED® test formula sheet. Just remember what the B and h are. It makes no difference which face you choose as the base in a rectangular prism because the height will always be the height of the third face, perpendicular to the base.

For example, for a $3 \times 4 \times 6$-inch rectangular prism, the surface area is

$$SA = 2[(3 \times 4) + (3 \times 6) + (4 \times 6)]$$

$$= 2(12 + 18 + 24) = 2(54) = 108 \text{ square inches.}$$

For the same rectangular prism, the volume is

$$V = (3)(4)(6) = 72 \text{ cubic inches.}$$

Example 7.15.

Priority mail shipping boxes at the post office are sold in specific sizes, such as the following (all are inside dimensions in inches):

$$12 \times 12 \times 5.5$$

$$5.38 \times 8.63 \times 1.63$$

$$11 \times 8.5 \times 5.5$$

a. What is the interior surface area of the biggest box?

b. What is the volume of the middle-size box?

Answer 7.15.

a. First, which is the biggest box? Since all three boxes have one side that is about the same (5.5 or 5.38 inches), compare only the other two dimensions. It is clear that 12×12 will be larger than either 11×8.5 or 5.38×8.63, so we don't have to calculate the sizes of the boxes, just do a comparison of the dimensions. The surface area of the largest box, with dimensions $12 \times 12 \times 5.5$, is $SA_{\text{rectangular prism}} = ph + 2B$, according to the GED® test formula sheet. B is the base area, so let's take the base as the largest side, $12 \times 12 = 144$ square inches. The perimeter of the base, since all edges are 12 inches, is $4 \times 12 = 48$ square inches. The height to the base is the remaining measure, 5.5 inches. Therefore, the surface area is

$$SA_{\text{rectangular prism}} = ph + 2B$$

$$SA = (48)(5.5) + 2(144) = 264 + 288 = 552 \text{ square inches.}$$

b. We can determine which of the remaining two boxes is the medium-size one by observation. Two of their dimensions are close to the same, but the third dimension is 11 inches for one box and 1.63 inches for the other. There is no doubt which of these is larger, so the medium-size box has interior dimensions in inches of $11 \times 8.5 \times 5.5$, and its volume is given by (see the GED® test formula sheet)

$$V_{\text{rectangular prism}} = lwh$$

$$V = 11 \times 8.5 \times 5.5 = 514.25 \text{ cubic inches.}$$

Cubes

Since a square is a rectangle with equal sides, a **cube**, which has twelve edges and six square faces (see the figure below), is included in this classification. So a cube is a rectangular prism and so is a cereal box. (But if you are thinking of oatmeal, it may come in a cylinder-shaped box, which is discussed in the next section.)

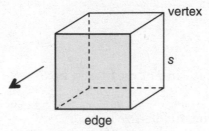

Therefore, the surface area of the cube is

$$SA_{cube} = 2s^2 + 2s^2 + 2s^2 = 6s^2.$$

The volume of a cube comes from the volume of a rectangular solid with $l = w = h = s$. So

$$V_{cube} = s^3.$$

Example 7.16.

Nina uses a plain wooden cube as a plant stand. She wants to cover it with a patterned adhesive paper, but first she must figure out how much paper she needs to buy.

a. If the cube is 18 inches on an edge and she wants to cover all the sides, how many square feet of paper does she need?

b. If the top of the cube is removable and the thickness of the cube is 1 inch all around, how much volume (in cubic inches) does Nina have to stash her magazines out of sight?

Answer 7.16.

a. 13.5. First of all, the paper is sold in square feet, so we should convert the 18 inches into 1.5 feet to make the calculation easier. Then the square footage Nina needs is $(1.5)^2 = 2.25$ square feet for each side times the number of sides in a cube (which is 6), or $6(2.25) = 13.5$ square feet of paper. The GED® test formula sheet shows $SA_{cube} = 2s^2 + 2s^2 + 2s^2 = 6s^2$, which is exactly what we did here.

b. 4096. The question asks about volume in cubic inches, so we can leave the dimension in inches. The inside "edge" of the cube will be the outside edge minus 1 inch on each side, or 18 − 2(1) = 16 inches as the inside measure. Then the volume (from the GED® test fomula sheet) is given by $V = s^3 = (16)^3 = 4096$ cubic inches.

Cylinder

A **cylinder** is in the shape of a can (or the usual container for oatmeal, as we said earlier). If you take the cylinder apart, you will see that it consists of two round ends plus a rectangle that wraps around the can shape.

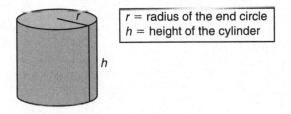

> r = radius of the end circle
> h = height of the cylinder

The *surface area* of a cylinder is the area of the two circular ends ($2\pi r^2$) plus the area of the rectangle for the "wrap." One side of this rectangle is the height of the cylinder, h, and the other side is the circumference of the circular end ($2\pi r$), so the area of the rectangle is $2\pi rh$.

Therefore,

$$SA_{\text{cylinder}} = 2\pi rh + 2\pi r^2,$$

which is the formula given on the GED® test formula sheet. Just remember what the r and h stand for.

HINT

Think of a soup can with the label peeled off. Then you can see that the ends are two circles, and the label is the area around the cylinder.

The *volume* of a cylinder follows the rule that it is the area of the base (πr^2) times the height, h, so

$$V_{\text{cylinder}} = \pi r^2 h,$$

which is the formula given on the GED® test formula sheet. Just remember what the r and h mean.

Example 7.17.

How much paper does Lisa need to put a label of her own design on a can that is 6 inches high and has a diameter of 4 inches? (Use $\pi = 3.14$. Round your answer to the nearest square inch.)

Answer 7.17.

75 square inches. The label on the can is the same as the surface area of the side of a cylinder, which is actually a rectangle with a length that equals the circumference of the can and a width that is the height of the can. The circumference of a circle (in this case, the end of the can) is $C = \pi d = 4\pi$. So the dimensions of the label will be $A_{rectangle} = lw = (4\pi)(6) = 24 \times 3.14 = 75.36$, which rounds to 75 square inches. Note that the formulas used in solving this problem are not on the GED® test formula sheet.

Pyramid

We will be dealing with a right regular **pyramid**, which is a three-dimensional figure in which one base is a regular shape (all the sides are equal), the other end is a point that is exactly over the middle of the base, and all of the other faces are identical triangles.

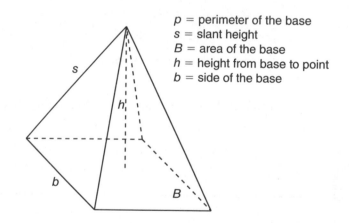

p = perimeter of the base
s = slant height
B = area of the base
h = height from base to point
b = side of the base

The *surface area* of a pyramid is the sum of the areas of the base and the triangles that form the sides. Obviously, if the base is a square, the sides are four triangles, and if the base is a triangle, three other triangles (not necessarily identical to the base triangle) form the sides.

A new measurement is introduced with figures that come to a point, the **slant height**, s, which is the measure from the point to the middle of the side of the base. Since the base is regular (all sides are equal), the slant heights for all sides of a pyramid are equal. The area of each triangle is $\frac{1}{2}bs$, where b is a side of the base of the pyramid, and s is the slant height. Since we have as many triangles as the number of sides (n) of the pyramid base, when we add all the triangle areas up, we get $n(\frac{1}{2}bs)$, but nb is the perimeter ($p = nb$) of the pyramid base. So the total area of all the sides is $n(\frac{1}{2}bs) = \frac{1}{2}nbs = \frac{1}{2}ps$. The area of the base is defined as B, and it is the area of whatever the base shape is, such as b^2 for a square. The final equation is thus

$$SA_{pyramid} = \frac{1}{2}ps + B,$$

which is the formula given on the GED® test formula sheet. Just remember what the B, p, and s are.

The *volume* of a pyramid, instead of being Bh, as volume was for the prism and cylinder, is now $\frac{1}{3}Bh$, where B is the area of the base and h is the perpendicular height from the base to the point. It is not the slant height. This factor of $\frac{1}{3}$ is used for all regular solids in which one "base" comes to a point. The reasoning behind using this factor is beyond the scope of the GED® test. The volume of a pyramid is

$$V_{pyramid} = \frac{1}{3}Bh,$$

which is the formula given on the GED® test formula sheet. Just remember what the B and h stand for.

Example 7.18.

The approximate measurements of the Great Pyramid of Giza in Egypt are:

Height = 480 feet

Sides = 755 feet at base

Slant height = 610 feet

a. What is the surface area (don't include the base) of the Great Pyramid of Giza in Egypt?

b. How many acres does the pyramid cover? (An acre is equal to 43,560 square feet, rounded to the nearest whole number.)

Answer 7.18.

a. 921,100 square feet. The surface area of a pyramid (from the GED® test formula sheet) is $SA_{pyramid} = \frac{1}{2}ps + B$, but the question asks for the surface area excluding the base (since we cannot see that). So the equation to use is just

$$SA_{Great\ Pyramid} = \frac{1}{2}ps,$$

where p is the perimeter of the base. Here the base is a square with side equal to 755 feet, so $p = 4(755) = 3020$ feet. The s in the pyramid equation is not a side, but the slant height, 610 feet. So

$$SA_{Great\ Pyramid} = \frac{1}{2}(3,020)(610)$$

$$= 921,100 \text{ square feet}$$

b. 13. Asking how much area the pyramid covers is the same as asking for the area of the base. Since the base is a square, the area is found by squaring one of the sides of the base, which are also the bases of the triangles, 755 feet each. So the area of the base is $(755)^2 = 570,025$ square feet. To find out how big this is in acres, use the proportion $\frac{1 \text{ acre}}{43,560} = \frac{x \text{ acres}}{570,025}$, or, by cross-multiplication, $43,560x = 570,025$, and $x = \frac{57,025}{43,560} = 13.09$, which rounds to 13 acres. Impressive!

Cone

A **cone** is indeed cone-shaped like an ice cream cone that comes to a point or a disposable cup that comes to a point. The measures of the surface area and volume are not as straightforward as they were for the other three-dimensional figures we have considered. It is important to know the meanings of the variables in the formulas given on the GED® test formula sheet, but we have encountered them in the discussion of the pyramid above. The only difference now is that the base is a circle and not a regular polygon.

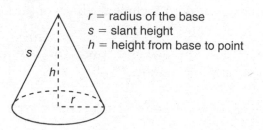

r = radius of the base
s = slant height
h = height from base to point

The surface area of a cone is the area of the base (circle) and the area of the "wraparound." This last area actually comes from the formula for the pyramid with a little imagination. Imagine that instead of, say, a square as the base of the pyramid, we have a regular polygon of 100 sides. The formula will still be the perimeter of the base times the slant height, right? Now imagine the base has too many sides to count—the base is coming very close to being a circle, and the perimeter is coming very close to being its circumference. So instead of the pyramid's surface area of $\frac{1}{2}ps + B$ that we found in the last section for the pyramid, we substitute the circumference ($2\pi r$) for the perimeter p. Also, B, the area of the base, now becomes the area of the circle of the cone (πr^2). So we get $SA_{cone} = \frac{1}{2}(2\pi r)s + \pi r^2$, which simplifies to

$$SA_{cone} = \pi rs + \pi r^2.$$

This formula is given on the GED® test formula sheet. Just remember what the r and s are.

The volume of a cone follows the same reasoning as for the pyramid, except that for the cone, the base is a circle of area πr^2. The factor of $\frac{1}{3}$ is used because the figure comes to a point. And the height is the perpendicular height h from the point to the circular base. So the volume of a cone is given by

$$V_{cone} = \frac{1}{3}\pi r^2 h,$$

which is the formula given on the GED® test formula sheet. Just remember what the r and h stand for.

Let's do a fun example now.

Example 7.19.

How many round candies that each have a volume of $.003\pi$ cubic inches will fill a sugar cone that is 7 inches long and has a diameter of 3 inches if space between the candies isn't taken into consideration? (If those round candies sound really small, they actually are in the form of little balls $\frac{1}{4}$ of an inch in diameter—not that tiny.)

Answer 7.19.

1,750. We have to figure out the volume of the sugar cone. The formula for the volume of a cone (from the GED® test formula sheet) is

$$V_{cone} = \frac{1}{3} \pi r^2 h.$$

We need the radius for this equation, but since we know the diameter of the cone, the radius is just half of it: $(\frac{1}{2})(3) = (\frac{3}{2})$. So we have

$$V_{sugar\ cone} = \frac{1}{3} \pi(\frac{3}{2})^2(7) = \frac{21}{4}\pi = 5.25\pi.$$

Therefore, the number of candies that will fit into the cone is $\frac{5.25\pi}{.003\pi} = 1,750$. Really?

Sphere

A **sphere** is a three-dimensional figure in which every point on the sphere is the same distance from a point called the **center** of the sphere. That distance is called the **radius** of the sphere, r. The idea is that it is like a three-dimensional circle.

A common example of a sphere is a ball. The reason spheres are so common in nature (a drop of water, for example, or the moon) is that they have the smallest surface area for a given volume. The formulas for the surface area and volume of a sphere have a factor of 4 in them that can be explained only by using calculus, so we will just accept the formulas as given.

The surface area of a sphere is similar to the area of a circle (πr^2) with a factor of 4:

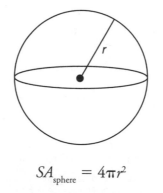

$$SA_{sphere} = 4\pi r^2$$

Likewise, the formula for the volume of a sphere is similar to the formula for the volume of a cone where the height is the radius and again we have the factor of 4:

$$V_{sphere} = \frac{4}{3} \pi r^3.$$

Remember, all the formulas for the three-dimensional figures are on the GED® test formula sheet.

Example 7.20.

A sphere is inside a cube so that it touches every side of the cube. If an edge of the cube is 10 inches, what is the difference between the volumes of the cube and the sphere? (Use $\pi = 3.14$ and round your answer to the nearest hundredth.)

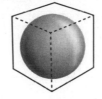

Answer 7.20.

476.67 cubic inches. The volume of the cube is $V_{cube} = s^3 = 1000$ cubic inches. The volume of the sphere is $V_{sphere} = \frac{4}{3}\pi r^3$, so we need to know the radius of the sphere. We know the diameter of the sphere because it has to be the same as the edge of the cube, or 10 inches. So the volume of the sphere is $V_{sphere} = \frac{4}{3}\pi(5)^3 = \frac{4}{3}(3.14)(125) = 523.33$ cubic inches. The difference is $1,000 - 523.33 = 476.67$ cubic inches.

Scaling

Scaling a geometric figure means multiplying all dimensions of the figure by the same factor, such as doubling the edge of a cube or tripling the diameter of a sphere. The least complicated three-dimensional figures to scale are **regular** solids, such as the cube, circle, or equilateral pyramid (one in which all sides and the base are equilateral triangles).

But what happens to the surface area or the volume when we scale a regular solid? The simple answer is that if the **scale factor** (amount the figure is scaled by) of the edge or diameter of a regular solid is x, the scale factor for the surface area is x^2, and the scale factor for the volume is x^3.

Imagine two cubes, one with an edge of 1 inch and the other with an edge of 3 inches. The scale factor between the edges of these two cubes is three. The area of any face of the larger cube is $3^2 = 9$ square inches, or nine times the area of any face of the smaller cube. Since volume involves the product of three dimensions (length $\times$ width $\times$ height), and each edge in the smaller cube is scaled by three in the larger cube, the volume is $3 \times 3 \times 3 = 3^3 = 27$ cubic inches, or 27 times the volume of the smaller cube.

HINT

If two three-dimensional figures that have exactly the same shape are scaled by a factor of x, their surface areas are scaled by a factor of x^2 and their volumes are scaled by a factor of x^3. Since areas are given in *square* units, you *square* the scale factor, and since volumes are given in *cubic* units, you *cube* the scale factor.

Example 7.21.

Quinoa (pronounced keen-wah) is a vegetable seed that has become popular as a substitute for rice because it is low-calorie and has a high fiber content. Raw quinoa is small—smaller than the size of the "o" in this typeface. For the purposes of this example, let's assume the seed is spherical (it almost is) and has a diameter of 2 mm and a volume of $\frac{4}{3}\pi$ mm^3 = 1.33π mm^3. When quinoa is cooked, its volume increases four times. What is the volume of one cooked quinoa seed? (Use π = 3.14 and round your answer to the nearest hundredth.)

Answer 7.21.

16.75 mm^3. The volume of the cooked quinoa is 4(1.33π) mm^3 = 16.71 mm^3. Before you think, "Wow! That's a huge seed," note that this converts to about .001 cubic inches. In other words, one thousand cooked quinoa can fit in a cube one inch on a side.

Composite Three-Dimensional Figures

Three-dimensional composite figures are made up of more than one solid. In three dimensions, this could be a solid that is composed of a pyramid on a square.

One of the most famous composite figures is the Washington Monument, which is an obelisk. An obelisk is a composite figure that has an upright, four-sided pillar, which looks like a gradually tapering pyramid with its top cut off, topped off by a smaller pyramid. The Washington Monument obelisk is 555.125 feet tall and is the tallest structure in Washington, D.C. The entire surface area of the monument is 93,408.82 square feet, and the surface area of the top pyramid section is 3,824.875 square feet.

For any three-dimensional composite figure, the surface area is found by finding the surface area of each part that "shows." For example, for the Washington Monument, it would include only the areas of the four trapezoids that make up the base of the obelisk plus the surface area of the top pyramid (without its base). In contrast, the volume of any three-dimensional composite figure is found by adding the volumes of each part.

Exercises

1. In right triangle ABC, $BC = \dfrac{3}{4} AC$, $BC = 6$.

   a. The area of $\triangle ABC$ is ▢ square units.

   b. The length of AB is ▢ units.

2. Which of the following rectangles has the same area? (The figures are not drawn to scale; use the given measurements.)

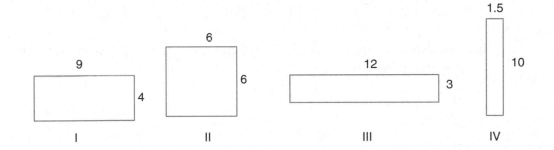

   A. II and III only

   B. I and III only

   C. I, II, and III only

   D. All of them

3. The figure below is made up of five squares, 3 units on a side. What is the perimeter of the figure?

   A. 36 units

   B. 24 units

   C. 16 units

   D. 45 units

4. How many minutes have passed when the hour hand of a clock has moved 60 degrees?

    A. 240 minutes

    B. 120 minutes

    C. 60 minutes

    D. 10 minutes

5. What is the area of isosceles triangle *PQR* inside circle *O* pictured below?

    A. 25

    B. 50

    C. 10π

    D. 25π

6. The volume of a cylinder is 48π cubic inches.

    a. If the height of the cylinder is 12 inches, what is the length of the diameter of the circular end in inches?

    A. 2

    B. 4

    C. 8

    D. 12

    b. If instead the cylinder were a cone with volume 48π and height 12 inches—what would be the length of the diameter of the base in inches?

    A. 12

    B. 4

    C. $2\sqrt{3}$

    D. $4\sqrt{3}$

7. In Circle *O*, the sides of triangle *ABC* are all integers and *AC* is the diameter.

    a. If the length of *OC* is 2.5 inches, what is the area of triangle *ABC*, in square inches?

    A. 3

    B. 4

    C. 6

    D. 12

b. If ∠BAC is 36.9°, what is the measure of ∠BCA?

    A. 53.1°

    B. 126.9°

    C. 36.9°

    D. cannot tell from the information given

8. A pipe has an outer diameter of 6 inches and inner diameter of 5 inches. The thickness of the pipe is ⬚ inch(es).

9. Figure *ABCDE* is a trapezoid with an equiangular triangle on top. If *BD* is 6 inches and *D* is the midpoint of *CE*, what is the perimeter of *ABCDE* ?

    A. 36 inches

    B. 36 square inches

    C. 12 inches

    D. Not enough information is given.

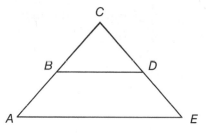

10. (Pull-down question.) Given the rectangle shown at the right, if the length is doubled and the width is halved, the area of the new rect-

angle is ⎧ double ⎫
       ⎨ the same as ⎬ the area of the original rectangle.
       ⎩ one-half of ⎭

11. A rug has an area of 64 square feet and a perimeter of 40 feet. Which of the following are the dimensions of the rug?

    A. 8 feet × 4 feet

    B. 4 feet × 16 feet

    C. 10 feet × 10 feet

    D. 8 feet × 8 feet

12. If three angles of a quadrilateral are 45°, 65°, and 130°, the remaining angle is ⬚ °.

13. Antonio knows that the area of his rectangular garden plot is 324 square feet. Now he wants to fence it in, but he isn't sure how much fencing to buy. The shorter side of the garden is 12 feet. How much fencing should he buy?

    A. 27 feet

    B. 39 feet

    C. 78 feet

    D. 324 feet

14. In the figure below, the small circle has a diameter of 6 inches. The area of circle O is

$$\left\{ \begin{array}{c} \text{twice} \\ \text{3 times} \\ \text{4 times} \end{array} \right\} \text{ the area of the smaller circle.}$$

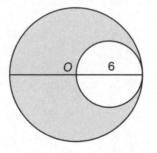

15. What is the radius of the smaller of two concentric spheres (one inside the other with the same center) if the ratio of the volumes of the two spheres is 8:1 and the larger circle has a radius of 4 inches?

    A. 1 inch

    B. 2 inches

    C. 3 inches

    D. 4 inches

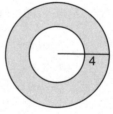

16. A cube and a rectangular prism have the same volume. If the sides of the rectangular prism are 4, 8, and 16, what is the length of a side of the cube?

    A. 2

    B. 8

    C. 64

    D. 512

Solutions

Answer 1. **a.** 24. If $BC = \frac{3}{4} AC$ and $BC = 6$, then $AC = BC \div \frac{3}{4} = 6 \div \frac{3}{4} = 6 \times \frac{4}{3} = 8$.

Then this is a right triangle with legs 6 and 8. Area $= \frac{1}{2} bh = \frac{1}{2} \times 6 \times 8 = 24$.

Remember the $\frac{1}{2}$ factor when finding the area of a triangle.

b. 10. From part (a), this is a right triangle with legs 6 and 8, so this is a 3-4-5 triangle (doubled), and the hypotenuse is 10. If you don't remember this, then you can use the Pythagorean formula to get the same result.

Answer 2. (C) I, II, and III only. The area of a rectangle is $A = lw$. For I, $A = 9 \times 4 = 36$; for II, $A = 6 \times 6 = 36$; for III, $A = 3 \times 12 = 36$; for IV, $A = 1.5 \times 10 = 15$. Therefore, the correct answer choice is (C).

Answer 3. (A) 36 units. Each outside square contributes three sides to the perimeter of the figure, so the perimeter is 4 (squares) $\times$ 3 (sides) $\times$ 3 (units on each side) = 36 units. Answer choice (D) is the correct number if the question had asked for the area of the figure (it would be 45 *square* units), but it asked for the perimeter.

Answer 4. (B) 120 minutes. Since there are 12 hours on a clock face and 360 degrees in a circle, each hour is represented by the hour hand moving 30 degrees. Therefore, 60 degrees represents 2 hours, or 120 minutes. Answer choice (D), 10, is wrong because it is the number of minutes when the *minute* hand moves 60 degrees.

Answer 5. (A) 25. The area of a triangle is $A = \frac{1}{2} bh$. The height of this triangle is $OQ = 5$, which is the radius of circle O. The base of this triangle is the diameter of circle O, so it must be $2(5) = 10$. Therefore, the area of $\triangle PQR$ is $A = \frac{1}{2}(10)(5) = 25$. Answer choices (C) and (D) should be eliminated right away since they contain π, and the question is about the area of a triangle, not a circle. In fact, (C) is the circumference of circle O, and (D) is the area of circle O.

Answer 6. **a.** (B) 4. The volume of a cylinder is given by $V_{cylinder} = \pi r^2 h$, so $48\pi = \pi r^2(12)$, and $r^2 = 4$. Therefore, $r = 2$, and the diameter is $2(2) = 4$.

b. (D) $4\sqrt{3}$. The volume of a cone is $V_{cone} = \frac{1}{3} \pi r^2 h$, so $48\pi = \frac{1}{3} \pi r^2(12)$, and $4r^2 = 48$, so $r^2 = 12$. Therefore, $r = \sqrt{12} = \sqrt{4} \times \sqrt{3} = 2\sqrt{3}$ and the diameter is $2(2\sqrt{3}) = 4\sqrt{3}$.

Answer 7.   **a. (C) 6.** First, we must find the sides of the triangle, which is a right triangle because it is inscribed in a semicircle. The hypotenuse is the diameter = 2(2.5) = 5. This must be a 3-4-5 right triangle. It is important that the problem says that all of the sides of the triangle are integers; otherwise, the legs of the triangle could be any two numbers as long as their sum was >5 and their difference was <5. So $AB = 4$ and $BC = 3$, and the area of a right triangle is one-half the product of the legs, or area = $\frac{1}{2}(4)(3) = 6$ square inches.

b. (A) 53.1°. Since triangle ABC is a right triangle, we know two of the angles, and the third angle is found by subtracting their sum from 180°, which is the sum of the angles in any triangle. Therefore, $\angle BCA = 180° - (90° + 36.9°) = 180° - 126.9° = 53.1°$.

Answer 8. $\frac{1}{2}$. The difference between the outer diameter and inner diameter is twice the thickness of the pipe, so the pipe is $\frac{1}{2}$ inch thick.

Answer 9.   **(A) 36 inches.** Even though triangle ACE looks like an equiangular triangle, we have to prove that it is. Then we can find the perimeter right away.

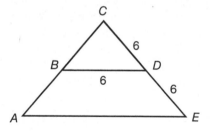

Triangle BCD is equiangular, so $\angle CBD = \angle CDB = 60°$; thus, $\angle BDE = \angle DBA = 120°$. Therefore, $\angle A = \angle E = 60°$, and triangle ACE is equiangular, which means it is also equilateral. Since BD is 6 inches, $BC = CD = 6$ inches also. Since D is the midpoint of CE, $CD = DE = 6$ inches too. Therefore, side $CE = 12$ inches, and the perimeter of figure $ABCDE$ is 3(12) = 36 inches. It probably took longer to read this explanation than to actually do the problem.

Answer 10.   The same as. The area of the original rectangle is $A = lw$. The area of the new rectangle is $A = (2l)(\frac{1}{2}w) = lw$.

Answer 11.   **(B) 4 feet × 16 feet.** Two answer choices can be eliminated right away because the area is 64 square feet, and only choices (B) and (D) have such dimensions. To choose the correct answer, take the other piece of information, which is that the perimeter is 40. This problem can also be solved by using simultaneous equations. Call the dimensions l and w. If the area is 64, then $lw = 64$. If the perimeter is 40, then $2(l + w) = 40$, or $l + w = 20$. Then $l = 20 - w$. Substitute this into the area equation, and $64 = (20 - w)w$, or $w^2 - 20w + 64 = 0$. This gives two roots, $w = 4$ or 16, which yields $l = 16$ or 4.

Answer 12. 120°. The sum of the angles in a quadrilateral is 360°, so the fourth angle is 360° − (45° + 65° + 130°) = 120°.

Answer 13. (C) 78 feet. First, Antonio needs to know the longer dimension of the garden. The area is 324 and one side is 12. Because $A = lw$, the formula he must use is 324 = 12l, or l = 324 ÷ 12 = 27 feet. Then the amount of fencing is the perimeter of the plot, or $P = 2(l + w) = 2(12 + 27) = 2(39) = 78$ feet.

Answer 14. Four times. The diameter of the smaller circle is the radius of circle O. The area of the smaller circle is $\pi(3^2) = 9\pi$. The area of larger circle O is $\pi(6^2) = 36\pi$, or four times the size of the smaller circle.

Answer 15. (B) 2 inches. The volumes of two three-dimensional objects with exactly the same shape, such as two spheres, varies as the cube of the linear dimension. Here that is the radius. So if the larger sphere has a volume that is 8 times the volume of the smaller sphere, the ratio is 8:1 for the volume and 2:1 (the cube roots) for the radii. Therefore, since the radius of the larger sphere is 4 inches, the proportion is $\dfrac{2}{1} = \dfrac{4}{x}$, where x is the radius of the smaller sphere, $x = 2$ inches.

Answer 16. (B) 8. The volume of the prism is $16 \times 8 \times 4$. For the side of the cube, we need to find $\sqrt[3]{16 \times 8 \times 4} = \sqrt[3]{512}$ because the volume of a cube is equal to a side cubed. You can find the value of $\sqrt[3]{512}$ using the virtual calculator. On the TI-30SX, just press 3, (2nd) (^), 512, and (enter).

List of Geometry Formulas Not on the GED® test Sheet

Figure	Perimeter	Area	Definitions
Triangle	$p = x + y + z$	$\dfrac{1}{2}bh$	x, y, and z are sides of the triangle. b is the base (any side). h is the perpendicular height to base b.
Square	$p = 4s$	s^2	s is the length of any side.
Rectangle	$p = 2(l + w)$	$A = lw$, or $A = bh$	l is the length and w is the width, or b and h are the length and width.
Circle	$C = \pi d = 2\pi r$	$A = \pi r^2$	C is the circumference, which is the perimeter of a circle; d is the diameter; and r is the radius (half the diameter).

Chapter 8

What Are the Chances?

Counting

The **basic counting principle** is one you probably already know. If something can be done in, let's say, three ways, A, B, and C, and something else can be done in let's say, two ways, D and E, the number of ways both can be done is six: A and D, A and E, B and D, B and E, C and A, C and E. Here we are assuming order doesn't make a difference—doing A and C is the same as doing C and A. This principle is the basis of a lot of what we do with data.

> **Basic Counting Principle:** If there are a ways for one activity to happen, and b ways for a second activity to happen, then there are $a \times b$ ways for both a and b to happen. This is true for any number of activities; for example, for five activities, this would be $a \times b \times c \times d \times e$.

An example or two will help you to remember this important principle. If a soft-serve shop has only vanilla but has two toppings, chocolate or caramel, you have two choices, right? Now suppose the shop has vanilla or chocolate soft-serve and offers an additional topping of chopped nuts. If you had to choose only one soft-serve and one topping, you would actually be able to choose from six combinations: vanilla with either chocolate, caramel, or nuts (3 choices) or chocolate with either chocolate, caramel, or nuts (3 more choices). Since you had 2 choices of soft-serve and 3 choices of toppings, you had $2 \times 3 = 6$ choices. And if the shop offered a third choice of swirl soft-serve and an additional topping of sprinkles, and you could choose only one soft-serve and one topping, then you would have $3 \times 4 = 12$ choices.

Now we will take this example one step further, but it will still be based on the basic counting principle. Let's say you could choose two toppings (but still only one soft-serve flavor). Then you would have 3 (for the soft-serve) × 4 (for the toppings) × 3 (for the additional topping) = 36 choices. Notice that the choice of an additional topping is now 3, not 4, because we aren't counting double caramel or anything like that—just two different topping choices.

You can imagine that different scenarios, such as allowing two flavors, would change the number of choices, but the idea is the same. Now we can go on to working with data.

Example 8.1.

Sally is packing for a 12-day trip. She wants to be able to mix and match her tops and pants so that she wears a different outfit every day but doesn't have to pack 12 tops and 12 pair of pants. Which of the following combinations will give her less to pack but still provide for 12 different outfits?

    A. 3 tops, 5 pants

    B. 4 tops, 4 pants

    C. 3 tops, 3 pants

    D. 3 tops, 4 pants

Answer 8.1.

(D) 3 tops, 4 pants. That will give her a choice of 3 × 4 = 12 outfits. The other answer choices will give her 15, 16, and 9 outfits.

Example 8.2.

A restaurant offers 3 types of meat, 2 types of potato, 4 types of vegetable, and 5 different desserts. How many meal combinations are there at that restaurant, assuming the patrons pick one of each category?

    A. 24

    B. 120

    C. 30

    D. 240

(B) 120. There are $3 \times 2 \times 4 \times 5 = 120$ choices.

Example 8.3.

What if, at the restaurant described above, there was an additional choice in each category: none. Then how many meals would there be? (Remember to subtract 1 from your calculation because someone who chose none in every category wouldn't have a meal, right?)

A. 359

B. 119

C. 59

D. 89

Answer 8.3.

(A) 359. Now there are $4 \times 3 \times 5 \times 6 = 360$ choices. Subtract the no-meal person, and the answer is 359.

Permutations and Combinations

Permutations and **combinations** tell us how many ways we can arrange things, like people sitting in a row, or books on a shelf, or the letters picked for a password. In this section we will discuss how many ways things can be arranged.

If we have, say, 5 people sitting in a row, the person who sits in the first seat could be picked from all 5 people, the next from 4 people, the next from 3, then from 2, and finally the last seat is a single choice. The counting principle tells us these choices can be made in $5 \times 4 \times 3 \times 2 \times 1 = 120$ ways. There is a word (and symbol) for multiplying from a number all the way down to 1, and that is **factorial**. The symbol is a little strange, it looks like an exclamation point. So we could have said that seating five people in a row can happen in $5! = 120$ ways.

The difference between permutations and combinations can be summed up in one word: *order*. In the example above, order made a difference. But if order didn't make a difference, there is only one way to pick five people from a group of five. Just pick them all and let them sit wherever they want. Where they sit will certainly differ, but there is still only one way to pick all five (out of a group of five) people. Then we can let them discuss where they will sit—whether Jason doesn't want to sit next to Shirley doesn't make a difference—we picked all of them in no particular order.

Permutations

In permutations, order matters, similar to using your password. If your password is BANK, and you enter the right letters but not in the right order (BNAK), your request will be denied. Another example of when order makes a difference includes picking three people from a group of seven members of a board of directors, where the first pick will be president, the second pick will be vice president, and the third pick will be treasurer. Thus, from the basic counting principle, and our intuition, we can figure out what a permutation is.

Let's look at this last example, picking three from seven when order makes a difference, and see whether we can get an idea of how to figure out how many ways this can be done. The first pick is one of 7 people, so there are 7 ways to pick the president. Then there are 6 remaining people, so 6 ways to pick the vice president, and finally 5 ways to pick the treasurer. According to the basic counting principle, then, the selection can be made in $7 \times 6 \times 5 = 210$ ways. The notation for this is $_nP_r$, which is read as picking r out of n things where order makes a difference. The importance of order makes it a permutation, thus the letter P.

There is a formula for finding the number of permutations, which is $_nP_r = \dfrac{n!}{(n-r)!}$. Let's see what happens when we use this. Just bear with this explanation because it has a happy ending—you don't have to do much math in the end. (We would have put an exclamation mark at the end of that last sentence, but we don't want you to think it was "end factorial.")

We have 3 people being picked from a group of 7, so $r = 3$ and $n = 7$. Then $_7P_3 = \dfrac{7!}{(7-3)!} = \dfrac{7!}{(4)!} = \dfrac{7 \times 6 \times 5 \times 4 \times 3 \times 2 \times 1}{4 \times 3 \times 2 \times 1}$. But look—the 4, 3, 2, and 1 all cancel, and we are left with $7 \times 6 \times 5 = 210$. In fact for all permutations, the number of multiplications of the factorial at the top reduces to just the first r factors. The rest will always factor out.

So if we pick 2 people out of 99 where the first gets the grand prize and the second gets the runner-up prize, we can do it in only 99×98 ways (multiplying just two factors). Or picking 2 colors out of a choice of 10 for the wedding tablecloths and napkins, in that order, can be done in—believe it or not—90 ways (that's 10×9).

Combinations

The number of combinations are smaller than permutations simply because order doesn't matter anymore. Take, for example, the group of 7 people we had earlier, when we picked three, and the first was the president, the second was the vice president, and the third was the treasurer. But now, let's consider a case where order doesn't make a difference. Let's look at the same seven-member board of directors, and we want three of them to serve on a committee. So we just pick three of them. Whether we pick Stacey, Dave, and DJ, or DJ, Dave, and Stacey, or any of the other four permutations of this group, it doesn't matter, as long as we have three people from the board on that committee. The original scenario, where the first pick was president and so on, had 210 permutations, but we now have only 35 ways to pick three people from a group of seven. We eliminated all the permutations that were the same committee (there were 175 of them!). So combinations are smaller than permutations because all the duplications are taken out.

The formula for combinations is $_nC_r = \dfrac{_nP_r}{r!}$, where the $r!$ removes all of the duplications, such as ABC being the same as BAC, BCA, etc.

HINT

An easy way to remember what makes permutations and combinations different:

Permutations arc like a **combination lock**. (What? Why not call it a permutation lock?) That is, which number is entered first, second, and third makes a difference. The lock won't open if you put the third number first. (Here we are talking about no repeats, such as 555.)

Combinations are like the **lottery**. For the plain Powerball lottery, five white balls are picked, but they can be in any order. If you got all of them, you really wouldn't care about the order anyway.

Example 8.4.

In how many ways can 3 of the 4 letters ABCD be rearranged if order makes a difference?

A. 12

B. 10

C. 24

D. 4

(C) 24. Since order makes a difference, this is a permutation. $_nP_r = 4 \times 3 \times 2 = 24$ ways.

Example 8.5.

How many ways can 3 of the 4 letters ABCD be picked in any arrangement?

    A. 12

    B. 10

    C. 24

    D. 4

Answer 8.5.

(D) 4. This is a combination, and we can list the possibilities: ABC, ABD, ADC, BDC. Any other combination is the same as one of these.

Probability

If something is sure to happen no matter what, we would say the chance it will happen is 100%, right? Likewise, if it can never happen no matter what, we would say the chance it will happen is zero. This is the basis behind **probability**.

Probability is expressed as a fraction, decimal, or percentage, which are all discussed in Chapter 3. As a fraction, it is a number from 0 to 1. As a decimal, it is a number from 0 to 1, corresponding to the 0% to 100% mentioned above. The GED® test will indicate which form is preferred for any particular problem.

As an example, consider a pair of dice, which are six-sided cubes with a number from 1 to 6 on each side.

If we tossed them as a pair, the probability that the total shown will be a number between 2 and 12 is 1, or 100%—the total must be some number between 2 and 12. What is the probability that the total is 13? That could never happen, so that probability is 0.

If something has an equal chance of happening as it does of not happening, such as getting heads on the flip of a coin, its probability is $\frac{1}{2}$, or we might say 50% or 50-50. But how do we calculate a probability that isn't one of these types of cases? By the simple reasoning behind it:

$$\text{Probability} = \Pr(x) = \frac{\text{number of successes}}{\text{number of possibilities}},$$

where $\Pr(x)$ means the probability of some event x happening.

But what do "successes" and "possibilities" actually mean? In the context of probability, success just means whatever you are interested in, and possibility means simply all the possibilities for success as well as failure. Simple examples are the best ways to understand probabilities for the GED® test because they are quite logical. After the examples, we will list all of the formulas you should know, but really they are all based on the definition of probability in the formula given above.

Suppose you have a jar containing 20 marbles. Eight of them are red, 7 are yellow, and 5 are blue. Now, if you choose a marble from the jar—without looking so it is a random pick—the following probabilities would be true:

$\Pr(\text{a marble is picked}) = 1 = 1.00 = 100\%$

$\Pr(\text{a red marble is picked}) = \dfrac{\text{number of red marbles}}{\text{number of marbles}} = \dfrac{8}{20} = \dfrac{2}{5} = .40 = 40\%$

$\Pr(\text{a green marble is picked}) = 0 = 0.00 = 0\%$

Then what would be the probability that a yellow marble is picked? You might have figured out that $\Pr(\text{a yellow marble is picked}) = \dfrac{7}{20} = .35 = 35\%$. That makes sense because there are fewer yellow marbles than red ones, so it should be a lower probability than for a red marble.

But how about the probability that it is red *or* yellow? That changes the number of successes to $8 + 7 = 15$, and $\Pr(\text{a red or a yellow marble is picked}) = \dfrac{15}{20} = \dfrac{3}{4} = .75 = 75\%$. Notice that this is the sum of $\Pr(\text{a red marble is picked}) + \Pr(\text{a yellow marble is picked}) = \dfrac{8}{20} + \dfrac{7}{20} = \dfrac{15}{20}$.

What if we wanted the probability that it is red *and* yellow? Then the probability would be 0, or 0%, because none of the marbles has two colors.

Let's now consider two picks and the probability of, say, picking a red marble on the first pick and a yellow one on the second pick. Suppose further that we replace the first marble in the jar before picking the second one. So Pr(red and yellow on two picks) = Pr(red) × Pr(yellow) = $\frac{2}{5} \times \frac{7}{20} = \frac{7}{50}$, or (.40)(.35) = .14, or 14%. What? So small? Yes, and it makes sense because on the second pick, you are starting with only a 40% chance of the desired series of picks. Did you notice that the answer would have been the same if we asked for the probability of picking a yellow marble on the first pick and a red one on the second pick?

This brings up an important aspect of successive picks: what's the difference if you replace the object (the marbles here) or not? If you replace the marble, there are still 20 marbles in the jar. Each pick **with replacement** has the same probability that it did when it was first picked.

Without replacement, though, both the number of successes and the number of possibilities change. After one marble is picked, there are only 19 marbles left, and there is one less of whatever color was picked the first time. So instead of having 8 red, 7 yellow, and 5 blue marbles, if the first pick was a red marble, then for the second pick, without replacement, Pr(a red marble is picked) is now $\frac{7}{19}$. The other probabilities change as well: Pr(a yellow marble is picked) = $\frac{7}{19}$, and Pr(a blue marble is picked) = $\frac{5}{19}$. And if a third marble is picked without replacement, the numbers (and probabilities) would change again.

Now, what if the choices weren't mutually exclusive? First of all, what does **mutually exclusive** mean, anyway? It means that there is no chance that one of the picks fits both criteria. It is the Pr(yellow and red) = 0 that we mentioned already—there is no chance in these examples for a marble to be both red and yellow. Let's consider another example to see what happens when the possibilities are not mutually exclusive.

Let's say we have a bowl of fruits and vegetables: 2 tomatoes, 4 bananas, 6 yellow squash, 8 red peppers, and 10 green peppers. Did we say this was a *big* bowl? It has to be. Now, let's look again at some probabilities when picking just one of them randomly:

$$Pr(fruit) = \frac{6}{30} = .20 = 20\%.$$

Wait a minute! Where did we get 6 fruits? An interesting fact is that *a tomato is a fruit, not a vegetable*. It has to do with how it grows (it is the fruit of the plant as opposed to lettuce, which is from the leaf of the plant and not the flower). It has nothing to do with how it is used, which is generally as a vegetable. Just don't put it in my fruit salad.

So $\qquad$ Pr(fruit or vegetable) = 100%.

$$\text{Pr(vegetable)} = \frac{24}{30} = .80 = 80\%.$$

$$\text{Pr(yellow)} = \frac{10}{30} = .\overline{33} = 33\%.$$

$$\text{Pr(yellow and a vegetable)} = \frac{6}{30} = .20 = 20\%.$$

What is the probability that something picked from the bowl is yellow or a vegetable? We said to add the probabilities in an "or" case, but if we do that here,

$$\text{Pr(yellow or a vegetable)} = \text{Pr(yellow)} + \text{Pr(vegetable)} = .80 + .33 = 1.13$$

But probabilities cannot be more than 1. What happened?

When we counted Pr(vegetable), we included the yellow squash. But when we counted Pr(yellow), we included the yellow squash again. So we counted the yellow squash twice—that's why .80 + .33 is greater than 1. We have to subtract one of the probabilities that it is a yellow squash from the product. In other words,

$$\text{Pr(yellow or a vegetable)} = [\text{Pr(vegetable)} + \text{Pr(yellow)}] - \text{Pr(yellow squash)}$$
$$= (.80 + .33) - .20 = .93$$

Let's check if this is true in our bowl. It's the probability of everything except the two tomatoes, our red *fruit*, so the probability by counting is $\frac{28}{30} = .93$. It works! This is because the two events, vegetable and yellow, weren't mutually exclusive. It was possible to have a yellow vegetable.

Example 8.6.

On an afternoon TV game show, contestants can choose one of ten prizes, numbered 1 through 10, sight unseen. Five of the prizes are almost worthless (for example, toilet bowl cleaner), two are worth $100, two are worth $1,000, and the grand prize is a new car. What is the chance that the ninth contestant will win the car?

A. $\dfrac{1}{2}$

B. $\dfrac{9}{10}$

C. $\dfrac{4}{5}$

D. $\dfrac{1}{10}$

Answer 8.6.

(D) $\dfrac{1}{10}$. The probability that the ninth contestant will win the car is the probability that no one before had won it, which is $\dfrac{9}{10} \times \dfrac{8}{9} \times \dfrac{7}{8} \times \dfrac{6}{7} \times \dfrac{5}{6} \times \dfrac{4}{5} \times \dfrac{3}{4} \times \dfrac{2}{3} \times \dfrac{1}{2} = \dfrac{1}{10}$ since all the numbers except the 10 cancel out. The probability is not just $\dfrac{1}{2}$ because the probability of the ninth person winning depends on the probabilities that all the previous contestants *didn't* win the car.

Data

The word **data** simply means "information," and it can have many forms: numbers, charts, tables, names, and political parties, just to name a few.

There are three basic types of data. Data that are **categorical**, as the name implies, relate to categories, such as the data shown in the following table, which reports how many people out of a group of 1,000 affiliate with each political party. As another example, answers to a "yes/no" survey provide categorical data.

Political Party Preference	Number
Democrat	368
Republican	390
Other/None	242
Total	1,000

Data that are **ordinal** (a word that comes from "order") relate to rank, such as answers to a survey that asks you to choose one of "strongly agree," "agree," "neither agree nor disagree," "disagree," or "strongly disagree." A form at the doctor's office that asks about your health—"poor," "reasonable," "good," "excellent"—also yields ordinal data. This can also be considered as a form of categorical data.

Categorical data are more descriptive (narrative) than quantitative data, but **quantitative** (or numerical) data give much more information about the **population** the data came from. *Population*, when it comes to working with data, doesn't mean the population of the United States, necessarily, but instead the population of interest. Quantitative data can tell us a lot about a particular population.

Data can be presented numerically (number of responses) in charts, graphs, tables, and even whimsical fashion (such as the *USA Today* Snapshots on the bottom of the front page of that newspaper). We will look next at how quantitative data tell us a lot about a specific population.

Central Tendency

What do we mean by **central tendency**? Just as the name implies, it tells us how the *center* of a group of data tends to look. It is the "typical" response for a set of data. It can be an average (the mean), the value of the data point in the exact middle of the data (the median), or even the value that has the greatest number of data points (the mode). The ways to determine these three measures of central tendency are not on the GED® test formula sheet, but you are expected to know them.

Mean

The **mean** of a group of numbers is the sum of the numbers divided by how many numbers there are. This formula comes right from the concept of average, and you probably already know it, although maybe not in these words.

$$\text{Mean} = \frac{\text{sum of data points}}{\text{number of data points}}$$

In school, for example, the usual measure of central tendency is the **average** grade, which is usually the mean of all the test scores.

We encounter averages all of the time in our daily lives: the average temperature for this day as told to us by the weather reporter on the news, the average lifespan of a U.S. male, and grade-point average. Now, when it comes to average temperature, we don't have *all* of the days for all time, so the weatherperson will probably say something like "since we started recording weather for this area." Likewise, we don't have data on *all* of the men in the United States, so we use a representative sample instead. (We will discuss samples later in this chapter when we talk more about statistics.) For a grade-point average, we probably have all of the data to come up with an accurate number.

Median

The **median** of a group of data is the value for which half the data are above and half are below. As the word *median* suggests (think of the median strip on a parkway), it is the *middle* number. To find the median of a group of data, the data must be put in order, either from lowest to highest (usual way) or from highest to lowest—it doesn't make a difference which way.

We just want to find the middle value. So we put the numbers in order and count how many data points there are. To get the *position* (not the value) of the middle number, we add one to the number of points and divide by 2. For example, the data set {2, 6, 4, 7, 2, 5, 8} in order becomes {2, 2, 4, 5, 6, 7, 8}. We can see that 5, in the fourth position, is the median. But if we didn't determine the position by sight, we say that there are 7 numbers, and $\frac{7+1}{2} = 4$. This tells us that the fourth position is the middle position. Notice that every number is listed, even repeat numbers, such as the 2 in this set.

For an *odd* number of data points, the median is an actual data point. For an *even* number of points, however, there are *two* middle numbers for which half the points are above them and half are below them. When you add 1 to an even number and divide by 2, you always get a "half" number, and that tells you the median is between two positions. Then the median is just the average (see above) of the numbers in these two positions. So for the data set {2, 6, 5, 8, 9, 10}, which is {2, 5, 6, 8, 9, 10} in order, there are six numbers, and the position of the median is $\frac{6+1}{2} = 3.5$, so it is between the third and fourth data points. Here that is between 6 and 8, so the median is the average of the third and fourth numbers, 6 and 8, $\frac{6+8}{2} = 7$. Thus, the median does not even have to be in the data set. It also does not have to be a whole number.

Again, the method for finding the median, which is just to find the middle number in a set of data points, is not on the GED® test formula sheet. You just have to remember how to find it, which you can from the name *median* (middle). Usually, you can just count the numbers to find the position of the median.

Mode

The **mode** is simply the most common value in the data set. So for the set {2, 4, 3, 6, 5, 6, 4, 6}, the mode is 6, the number that appears most frequently. Often it helps to put numerical data in order, but it isn't necessary. This is one measure of central tendency that doesn't have a formula associated with it. Since mode also is not on the GED® test formula sheet but you still have to know it, you should just remember that *mode* means "most frequent."

It is important to remember that it is the actual data point, not the frequency, that is the mode. In other words, for the data set above, the mode is 6, not 3 (the number of times 6 appears). Also, data can be bimodal (have two modes), as in {1, 1, 1, 2, 3, 3, 3, 4, 5, 6, 7, 8}, where 1 and 3 both have the same frequency, three.

For categorical data, this is the only measure of central tendency. For example, in the table on political preference, the mode is Republican (not the number 390) because there are more Republicans than either of the other two categories. We cannot have a mean for these data (what would it be?), nor a median since we wouldn't know how to "order" Democrat, Republican, and Other/None.

Skewed Data

Data that are **skewed** have extreme scores at one end of the ordered data that make the mean seem out of line with what the data actually look like. In those cases, the median is usually a better measure of central tendency.

For example, for the following data points,

2   3   4   4   5   6   6   6   7   8   9   10   80,

the mean is $\dfrac{2 + 3 + 4 + 4 + 5 + 6 + 6 + 6 + 7 + 8 + 9 + 10 + 80}{13} = 11.5.$

The mean is pulled in direction of the extreme value of 80. A more representative measure for clearly skewed data is the median (here it is 6).

Data reports can be manipulated to favor a particular viewpoint. Let's say that a labor contract is up for renewal, and the firm has collected the following data:

Mode = $62,000

Median = $78,000

Mean = $97,000

The reason the mean is so much higher than the other two measures is that a few people are making more than $200,000 per year. This inflates the mean but doesn't have an effect on the median. The union reports that the typical worker makes $62,000 (based on the mode) or that half of the workers make $78,000 or less (based on the median, which is the middle value in the data).

In contrast, management reports that the average salary is $97,000. Neither report is incorrect—it's all in what the emphasis is.

As another example, let's consider a small company with six employees making $45,000, $52,000, $59,000, $53,000, $196,000, and $45,000. The mean of these salaries is

$$\frac{45,000 + 45,000 + 52,000 + 53,000 + 59,000 + 196,000}{6} = 75,000$$

and the median is $52,500 (the salaries in order from lowest to highest appear in the mean calculation above). To create an image of success, the company's owner reports the mean rather than the median.

Since all figures are thousands of dollars, the calculation for this mean could have been streamlined to

$$\frac{45 + 45 + 52 + 53 + 59 + 196}{6} = 75$$

but we have to remember that this represents thousands of dollars, so the answer is $75,000.

Example 8.7.

Marisha took three tests and scored 70, 75, and 92. What is her average?

A. 70

B. 75

C. 79

D. 89

Answer 8.7.

(C) 79. We should expect her average to be higher than 75. Indeed, it is: $\dfrac{70 + 75 + 92}{3} = 79$.

Example 8.8.

Marisha is taking a fourth test. Her scores on the first three are 70, 75, and 92. She needs a score of [   ] on the fourth test to raise her average to 83.

Answer 8.8.

95. Now the problem changes slightly from what was asked in Example 8.1. Marisha has four, not three, tests, and we know her desired average but not the fourth test score. We use the same basic equation for the mean, but with the new information: $\dfrac{70 + 75 + 92 + x}{4} = 83$. Multiplying both sides by 4, we get $237 + x = 332$, or $x = 95$. So Marisha needs a score of 95 to raise her average to 83.

Example 8.9.

Now it's finals time for Marisha. The teacher counts the final exam score as the same as three test scores. Marisha scores a 90 on the final. Assuming her four other test scores are 70, 75, 92, and 95, what is her grade for the semester after the final?

    A. 80

    B. 83

    C. 86

    D. 92

Answer 8.9.

(C) 86. The fact that the final exam score counts the same as three test scores means the final exam score is **weighted**. The way to figure Marisha's grade is basically the same as before, except we have to count the final three times: $\dfrac{70 + 75 + 92 + 95 + 90 + 90 + 90}{7} = 86$. Based on the preceding two examples, answer choices (A) (B), and (D) should be eliminated because we know Marisha went into the final exam with an 83 average and that she did better than that (90) on the final, so her semester score has to be more than 83 and less than 90.

Example 8.10.

What is the median of the following data?

3   80   57   9   45   35   5   2   15   15   7   3   15   8   10   6   10   4   20   5

Answer 8.10.

Arrange the numbers from low to high:

2   3   3   4   5   5   6   7   8   9   10   10   15   15   15   20   35   45   57   80

There are 20 numbers, and the middle value is between the 10th and 11th positions, or between data points 9 and 10; it is 9.5. Again the median need not be a data point.

Spread of Data

Another description of a data set is how much the data are spread out. The **range** is simply the highest value minus the lowest value in a data set. For the data above, the range is 78 because the data go from 2 to 80, and $80 - 2 = 78$. It is that simple, highest minus lowest. (Do not get this range mixed up with the range of a function, which was discussed in Chapter 6.)

For an example of how the mean and range describe data, look at the following two sets of data:

A: {14, 15, 16, 17, 18},   B: {2, 2, 15, 16, 45}.

Let's figure the means for both sets.

$$\text{Mean}_A = \frac{14 + 15 + 16 + 17 + 18}{5} = 16,$$

$$\text{Mean}_B = \frac{2 + 2 + 15 + 16 + 45}{5} = 16.$$

The means are the same, but the sets are certainly different.

The range of set A is $\text{range}_A = 4$, whereas the range for set B is $\text{range}_B = 43$. Set B is far more spread out than set A. In fact, there is a measure called the **deviation from the mean**, and it is far greater for set B than it is for set A, indicating that the data points in set A are much closer to the center of the data. Although you don't have to calculate the deviation from the mean on the GED® test, you should be aware that there is a measure that tells the closeness to the center of a set of data. It is used quite a bit in statistics.

Statistics

Statistics are a way of presenting information about a set of data. Data can be described by their measures of central tendency as well as how much they are spread out from the center. Statistics are presented to the public in many forms: verbally, through tables, and through charts and graphs, among others.

We see statistics every day. The news reports the Dow Jones average. High schools report the percentage of incoming freshmen they expect to graduate. And at election time we are told what percentage of women, for instance, would probably vote for a particular candidate.

To understand the math behind statistics, we start with a **population** of interest. A population could be all the workers at a particular company, all the dogs who won or placed at the AKC championships, all the cans of chili con carne in a particular shipment, and so forth. In a statistical sense, population doesn't have to be people, but it does have to be all the "things" we are considering for the statistic.

Of course, logistically, we cannot query all the people in the United States, if that is our population of interest (for example, to determine statistics about hours of TV watched per day). For this reason, statisticians use a representative sample and then make assumptions about the whole population from that sample. Note, though, that the U.S. Census is an example of polling (or surveying) every unit in a population.

Notice the word *representative* here. For example, if we wanted to know about hours of sleep per night for seniors, we wouldn't sample teenagers; if we wanted to know about what toothpaste more dentists recommend, we wouldn't sample obstetricians. Data also have to be unbiased. This translates to having a **random sample**, which means everyone in the population has an equal chance of being picked for the sample.

As an example, to calculate people's opinions of a particular store, it might be okay to sample people as they leave that store, but not okay to sample only those in line to return merchandise, who are probably less-than-satisfied customers. Likewise, a telephone survey would not be appropriate for a survey of, say, all the people in Denver, even if only Denver phone numbers were used, because many people don't have landlines anymore, many people screen their calls, etc., so that sample would be biased only toward a segment of the population and would exclude everyone else. In addition, the timing of a survey can produce a biased sample. If a telephone survey is taken in the middle of the day, students and the working population are usually not included; if the survey is taken too late at night, however, the older population may not be included. No one in the population should be excluded from being in a sample in an unbiased statistical study.

We will not go into how statistics are mathematically calculated here, as that is beyond the scope of the GED® test and in fact involves higher math. But the principles, presentation, and uses of statistics are important to know in everyday life as well as on the GED® test. As we said, statistics can be presented in many forms.

Quantitative data are actual numerical data points that can be represented as lists, tables, histograms, frequency curves, and box plots. *Categorical data* are counts or percentages of individuals in categories and can be presented in the form of a table, but are best represented pictorially as, for example, in a pie chart (also called a circle graph) or a histogram. Scatter plots (also called dot plots) are used to show correlation between two variables, such as height and weight. You should be able to interpret any of these forms of presenting statistical data.

Tables for Quantitative Data

Tables are usually presented with one column for the item being measured and/or another column with the count (frequency) or percentage of the whole. For *quantitative* data, the measures of central tendency are often shown clearly in a table. As an example, let's look at the number of televisions in 30 randomly selected households:

1 3 3 0 3 1 1 2 2 1 6 2 3 1 1 2 3 2 3 3 2 2 2 1 2 4 2 3 3 1

These data are far more understandable when put in table form:

Number of Televisions	Number of Households
0	1
1	8
2	10
3	9
4	1
5	0
6	1

So we see that the majority of households have 1 to 3 televisions. Looking at the table, it is not surprising that the median and mode are both 2 televisions per household, and the mean is 2.17 televisions per household.

Sometimes, listing the data in table form wouldn't be much better for analysis than just listing the data as a string of numbers. By lumping together the data in a meaningful way, however, the data become more understandable. Consider the following data showing years of experience of 20 technicians in a lab:

15 8 21 17 12 27 14 16 4 2 18 34 2 3 7 9 11 5 2 2

If we put this in a table, it would have 17 entries, all with one response (except for the data for 2 years, which has four responses). But if we lump together five years of data in a group, the data become more meaningful. Remember that every other group in the table must also contain data for five years.

Years of Experience	Number of Technicians	Percentage of Total
1–5	7	35%
6–10	3	15%
11–15	5	25%
16–20	2	10%
21–25	1	5%
26–30	1	5%
31–35	1	5%
Total	20	100%

The information that we can gather from this table includes that most of the technicians have less experience, the newest technicians make up 35% of the workforce, and that technicians usually leave the workforce after 20 years of experience, assuming these data have been about the same for several years. Even though it looks like that is the case, we cannot deduce that the workforce is younger because we don't have data about the ages of the respondents. (It is possible that all technicians can be 55 years of age, but their years of experience vary.)

Histograms

Histograms are a way of showing the distribution of data. The horizontal axis shows the data in equal-size groups and the vertical axis shows the frequency of each group. Since the group intervals are equal (each is 5 years), the area of each bar in a histogram provides a visual cue for the weight of that group compared to others. The histogram for the data shown in the above table is shown below as a histogram.

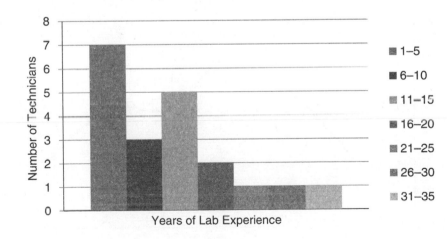

Histograms can have several formats. Another histogram format, this time showing how many days until a sample of 40 short-term bonds mature, is shown below.

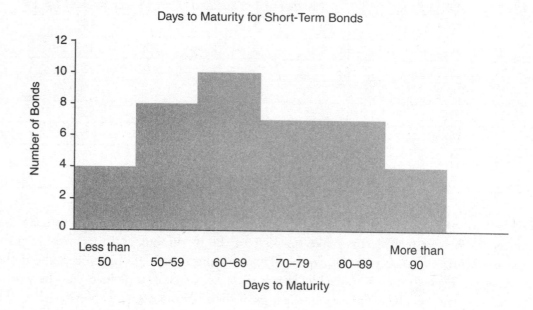

Frequency Curves

Frequency curves are similar to histograms in that they plot data versus frequency. If grouping is used, the point to be graphed occurs in the middle of the group. The frequency curve for the data from the histogram above is shown below. A frequency curve is a smooth line that connects the data points at the centers of the intervals.

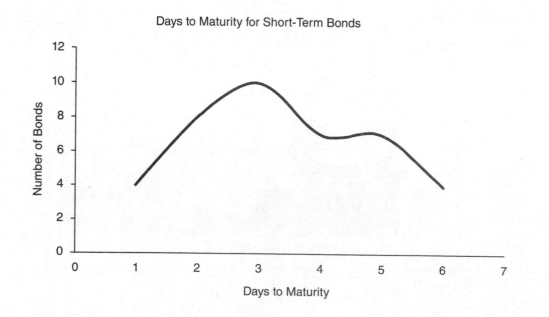

Boxplot

Boxplots provide ready data on the center (median) and variation in a data set. Box plots show quartiles. Just as the median (called Q2 in boxplot language) is at the 50% mark (half the data are above that value and half are below that value), the first quartile (Q1) is at the 25% mark (one-fourth of the data are below that value and three-fourths are above it), and the third quartile (Q3) is at the 75% mark (three-fourths of the data are below that value and one-fourth are above it). Q1 is actually the median of the values below Q2, and Q3 is actually the median of the values above Q2. Constructing a box plot isn't difficult, but it does take some time. On the GED® test, it is important that you be able to get information from a given boxplot.

For example, the boxplot shown below is for the following data set:

15 16 20 20 24 26 27 30 30 31 32 32 34 35 38 40 66

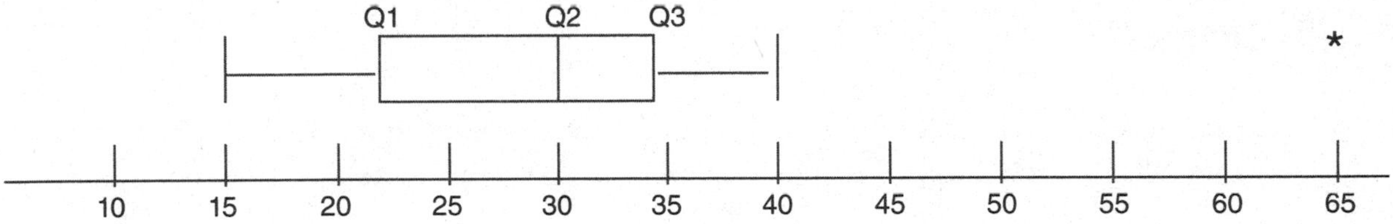

Here, Q1 = 22, Q2 = 30, and Q3 = 34.5. The asterisk at the far right of the box plot indicates an **outlier**. Looking at the data, you can see that 66 appears to be far outside the rest of the data, which defines an outlier. The two lines outside the box are at the lowest value of the data (15) and highest value of the data (40), not counting the outlier.

You should be able to read the following values from a boxplot (which may not have the Q1, Q2, and Q3 positions labeled), even though the actual data points may not be displayed:

lowest value

first quartile

second quartile (median)

third quartile

highest value

outlier

Table for Categorical Data

The table shown earlier for political party preference (reproduced here) is an example of a *categorical* data table.

Political Party Preference	Frequency
Democrat	368
Republican	390
Other/None	242
Total	1,000

This table is quite understandable and clearly better than a list of 1,000 responses. To make categorical data more manageable, it is best to use fewer than 12 categories, although if the categories have a purpose, such as comparing results for the 50 U.S. states, more categories may be used. Categorical data are shown quite well in pie charts and bar graphs.

Pie Charts

A **pie chart** is a circle divided into wedges that are proportional to the numbers in the data set. It is sometimes called a **circle graph**. The following pie chart shows the categorical data from the table above.

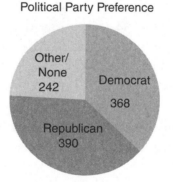

Political Party Preference

If the percentages of each category add up to more than 100%, pie charts will not work. An example includes a survey that says "Choose all that apply," such as sources of news, because some people could choose the newspaper as well as an online source or television. Another example is the language spoken in the home—some households may speak more than one language.

Bar Graph

A **bar graph** shows the categories on the horizontal axis and the frequencies on the vertical axis. This looks similar to a histogram, but the bars on a bar graph do not touch each other. The following bar graph shows the categorical data from the table above.

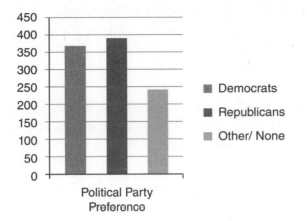

Scatter Plot

Scatter plots (also called **dot plots**) plot one variable against another to see whether there is a **correlation** (or association) between them. Scatter plots look similar to line graphs in that data points are plotted on horizontal and vertical axes. However, the purpose of the scatter plot is to show how much one variable is affected by another. Unless the two variables are perfectly correlated, the points will not form an exact straight line. The closer the points approximate a straight line (or a distinct curved line), however, the more correlated the variables are. Scatter plots must have a lot of data points on them to be effective in showing a trend.

The following scatter plot shows a good positive correlation between hours of study time (x-axis) and score on a 300-point test (y-axis). If there is a good correlation in data, the scatter plot may be used for prediction. For example, from the scatter plot shown below, we can predict that if someone studied for only 15 hours, her or his score would probably be around 125. The line through the points is the **line of best fit**. It doesn't go through all the points, but it shows the trend. Actually, the sum of the distances of the points above the line equals the sum of the distances of the points below it.

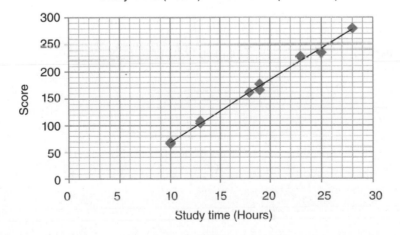

It is also possible to have a negative correlation, in which case the slope of the line would be from upper left to lower right, the same as the negative slope of a straight line. An example of negatively correlated data might be the age and price of a used car (as the age increases, the price decreases).

A scatter plot with no correlation has data points scattered randomly around the grid, and it looks like you can't make heads or tails of what is being shown. Sometimes a scatter plot will show two distinct groupings of data. A popular example of two distinct groups of data would be a scatter plot of time of day versus eruption of Old Faithful Geyser, which erupts several times a day.

Other Presentations

As with all data presentations, the method of presentation can be the standard ones discussed above, but they can be innovative as well. The following pie chart is presented as slices of a pie, in keeping with the data being presented. This is adapted from a *USA Today* Snapshot.

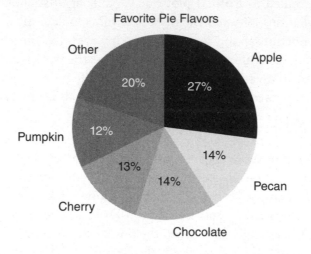

Favorite Pie Flavors

Example 8.11.

From the following boxplot, the value of the first quartile is ⬚ and the value of the third quartile is ⬚ . The range of the data is ⬚ .

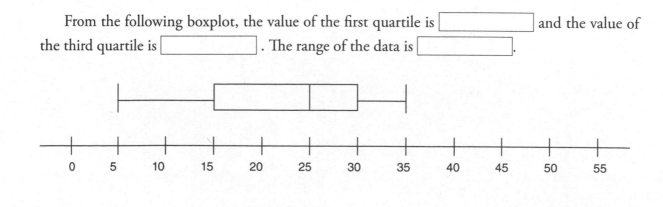

Answer 8.11.

15, 30, 30. The value of the first quartile is the left side of the box, or 15. The third quartile is the right side of the box, or 30. The range is the highest value (35) minus the lowest value (5), or $35 - 5 = 30$.

Example 8.12.

A census of 100,000 housing units in a city yielded the following table:

Number of Rooms in Unit	Number of Housing Units
1	515
2	1,753
3	6,890
4	22,672
5	30,624
6	27,338
7+	10,208

a. If a housing unit is chosen at random, the probability that it has exactly 6 rooms is

[   ] (round your answer to the nearest percentage).

b. If a housing unit is chosen at random, what is the probability that it has at least 6 rooms?

A. 27%

B. 10%

C. 37%

D. 38%

c. If a housing unit is chosen at random, what is the probability that it has at most 6 rooms?

A. 38%

B. 10%

C. 90%

D. 100%

a. 27%. It is $\dfrac{27,338}{100,000} = 27.338\%$, which rounds to 27%. The total of 100,000 was mentioned in the text above the table.

b. (D) 38%. The probability is $\dfrac{27,338 + 10,208}{100,000} = \dfrac{37,546}{100,000} = 37,546\%$, which rounds up to 38%. Note that "at least 6" means 6 or more.

c. (C) 90%. At most 6 rooms means everything from 6 rooms or less. On this chart, that would be everything except 7+ rooms, which has a probability of 10%. Therefore, the probability of at most 6 rooms is 100% − 10% = 90%.

Exercises

1. On a radio show, the fifth caller can choose a number, 1 through 10, representing ten prizes. Five of the prizes are almost worthless, two are worth $1,000, two are worth $5,000, and the grand prize is a trip around the world. Fill in the correct decimal answers.

    a. The probability that the caller will win the trip around the world is ⬚.

    b. The probability that the caller will win at least $1,000 is ⬚.

    c. The probability that the caller will win at most $1,000 is ⬚.

2. Makena's mom gave lollipops to each of the 12 children at the playground. She had 6 red lollipops, 3 blue, 2 green, and 1 yellow lollipop. The probability that a child received a lollipop other than green is

    A. $\dfrac{1}{6}$

    B. $\dfrac{2}{3}$

    C. $\dfrac{5}{6}$

    D. $\dfrac{3}{4}$

3. If factorial (designated by !) means the product of the integers down to one, 5! has the value ⬚.

4. The education levels of the chief executive officers (CEOs) of the 500 top U.S. companies are summarized in the following table.

Education Level	Number of CEOs
No college	15
Bachelor's degree	165
MBA	189
Law degree	52
Other higher education degree	79

a. How many of these CEOs have at least some college education?

A. 15

B. 79

C. 485

D. 165

b. How many have either an MBA or a law degree?

A. 241

B. 189

C. 52

D. 320

5. (No calculator allowed.) Three people from a group of five men and three women are chosen at random for an elite committee. The probability that none of the three committee members are women is [ ] . (Put your answer in the form of a fraction.)

6. Ten slips of paper numbered 1 to 10 are placed in a box. What is the probability of selecting an odd number and then an even number with no replacement?

A. $\dfrac{1}{2}$

B. $\dfrac{1}{4}$

C. $\dfrac{5}{18}$

D. $\dfrac{1}{5}$

7. For which of the following box plots is the median equal to half the sum of the values of the first and third quartiles?

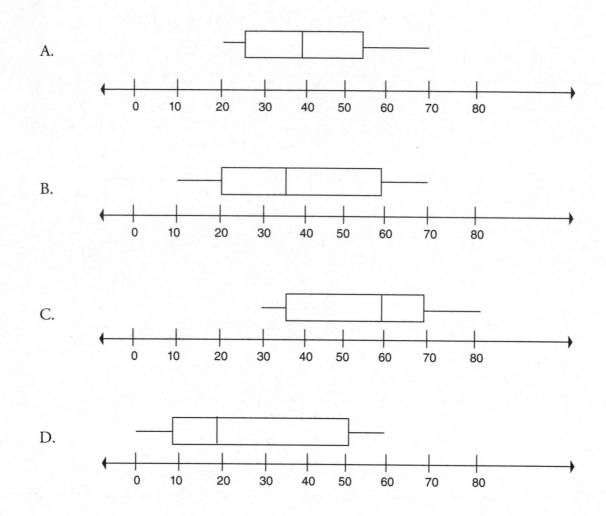

A.

B.

C.

D.

8. According to the World Trade Organization, in 2011 the world's largest exporters of clothing (in billions of U.S. dollars) were as shown on the following bar graph.

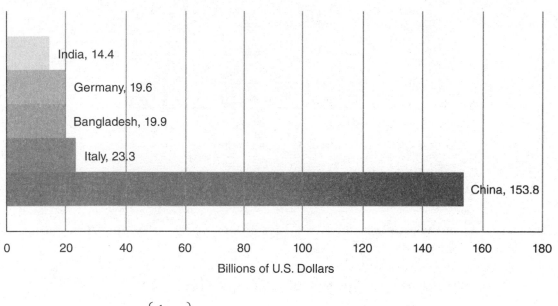

Largest Exporters of Clothing

China exports about $\left\{\begin{array}{l} \text{three} \\ \text{two} \\ \text{ten} \end{array}\right\}$ times as much as the other four countries combined.

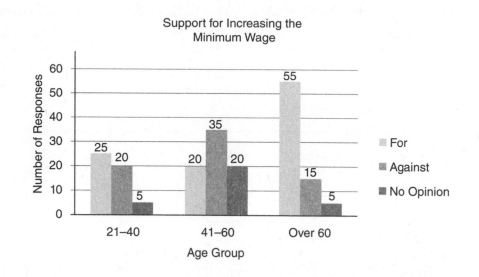

Support for Increasing the Minimum Wage

9. According to the chart above, which is based on a public opinion survey of 200 people,

    a. of the people in the 21−40 age group, [＿＿＿] % support increasing the minimum wage.

    b. which age group(s) had the most respondents?

    A. 21−40 age group

    B. 41−60 age group

    C. over 60 age group

    D. both the 41−60 age group and over 60 age group

10. Which data set simultaneously satisfies the following requirements:

    mean = 8, median = 9, mode = 2, range = 15

    A. {2, 2, 8, 11, 17}

    B. {2, 3, 9, 11, 17}

    C. {2, 2, 9, 10, 17}

    D {2, 8, 10, 13, 17}

11. Measures of central tendency include:

    A. mean

    B. median

    C. mode

    D. all of these

12. Categorical data can have

    A. a mean.

    B. a median.

    C. a mode.

    D. all of these.

13. Bar graphs and histograms $\left\{ \begin{array}{l} \text{are the same thing} \\ \text{have different uses} \\ \text{are used for quantitative data only} \end{array} \right\}$.

14. It $\left\{ \begin{array}{l} \text{is possible} \\ \text{is not possible} \\ \text{is unusual} \end{array} \right\}$ for the mode to be different than one of the data points.

15. An example of ordinal (ranked) data is

    A. the heights of the students in a class.

    B. success and failure.

    C. all the whole numbers between 1 and 2.

    D. strongly agree, agree, no opinion, disagree, strongly disagree.

Solutions

Answer 1. a. .10. $\frac{1}{10} = .10$

b. .50. $\frac{5}{10} = .50$

c. .70. $\frac{7}{10} = .70$

Answer 2. (C) $\frac{5}{6}$. There are two ways to do this problem. The first is to add up all the other lollipops (red, blue, yellow). The denominator is all of the lollipops, 12. Then the probability of not receiving a green lollipop is $\frac{6+3+1}{12} = \frac{10}{12} = \frac{5}{6}$. The other way is to figure out the probability of getting a green lollipop, $\frac{2}{12} = \frac{1}{6}$, and then the probability of not getting a green lollipop is $1 - \frac{1}{6} = \frac{5}{6}$.

Answer 3. 120. $5! = 5 \times 4 \times 3 \times 2 \times 1 = 120$.

Answer 4. a. (C) 485. "At least some" means everybody on this chart except the "no college" people, who number 15 out of 500 (the number 500 is in the text above the table, so you don't have to add up all the numbers). That total is $500 - 15 = 485$.

b. (A) 241. The question asks specifically about two degrees, the MBA (189) and law degree (52), so the answer is the sum of the two, which is 241. We are assuming here that no one has both an MBA and a law degree because then the data wouldn't be mutually exclusive. Answer choice (D) includes other higher education degrees, but the question didn't ask about them.

Answer 5. $\frac{5}{28}$. The probability that the first pick is a man is $\frac{5}{8}$. The probability that the second pick is a man is $\frac{4}{7}$. The probability that the third pick is a man is $\frac{3}{6}$. Each of these probabilities takes into consideration that the number of people decreases by one each time, as does the number of men. Therefore, the answer is $\frac{5}{8} \times \frac{4}{7} \times \frac{3}{6} = \frac{5}{28}$, after some cancellations.

Answer 6. (C) $\frac{5}{18}$. The probability of selecting an odd number is $\frac{5}{10}$. The probability of then selecting an even number is $\frac{5}{9}$. The combined probability is thus $\frac{5}{10} \times \frac{5}{9} = \frac{5}{18}$.

Answer 7. (A). When the median is in the exact middle of the box plot, it is the average of the other two measures, the first and third quartiles. Therefore, it is equal to half the sum of the values of the first and third quartiles. The median isn't in the middle of the box in the other answer choices.

Answer 8. Two. The total for the other four countries is US$77.2 billion, and China exports US$153.8 billion.

Answer 9. a. 50. The 21−40 age group has 50 respondents, and 25 of those are for increasing the minimum wage, which is $\frac{25}{50}$ = 50% of the group. Don't include the percentage mark (%) because it is already supplied.

b. (D) Both the 41−60 age group and over 60 age group. Both groups had 75 respondents, whereas the 21−40 group had 50 respondents.

Answer 10. (C) {2, 2, 9, 10, 17}. All choices have the correct range of 15. Eliminate answer choices (A) and (D) right away because the medians are not equal to 9. Then eliminate answer choice (B) either because the mode isn't 2 or because the mean won't be a whole number.

Answer 11. (D) All of these. The measures of central tendency are the mean, median, and mode.

Answer 12. (C) A mode. Categorical data are data gathered regarding categories. Think of responses that include "yes, no, or no opinion," which cannot be put in any order. These types of answers cannot have a median. The data do not have a numerical value, so they cannot have a mean. However, they can describe which category had the most responses, so their only measure of central tendency is the mode.

Answer 13. Have different uses. Although bar graphs and histograms look similar, bar graphs are used for categorical data, whereas histograms are used for numerical data.

Answer 14. Is not possible. In fact, since the mode reports the data point that has the most responses, it *must* be one of the data points.

Answer 15. **(D)** Strongly agree, agree, no opinion, disagree, strongly disagree. Rank order is the basis for ordinal data. For answer choice (A), the data for heights are not categories, but they are continuous numbers; the data are numerical. For answer choice (B), success and failure are categorical but not ranked data. Answer choice (C) has no data points.

GED® MATH REASONING

Practice Test 1

The GED® Math Reasoning test has five types of questions: multiple-choice, drop-down, fill-in-the-blank, hot-spot, and drag-and-drop. We represent all these question types in our practice tests. Please refer to the question-type descriptions starting on page 4 to refresh your memory of the question types on this test.

1. (No calculator allowed.) If David purchased x t-shirts at \$8 apiece, and y sweaters at \$30 apiece, which expression represents the total value of the purchases?

    (A) $8x - 30y$

    (B) $8x + 30y$

    (C) $30x - 8y$

    (D) $30x + 8y$

2. (No calculator allowed.) Indicate on the number line below the value(s) of x for which $\dfrac{-2+x}{x(x+2)}$ is undefined.

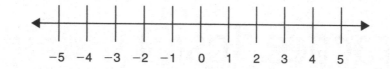

3. (No calculator allowed.) The commutative property, which says that order of doing an operation doesn't matter, allows us to say, for example, that $2 + 1 = 1 + 2$. The commutative property doesn't work for:

  I. subtraction

  II. multiplication

  III. division

    (A) I and II

    (B) I and III

    (C) I, II, and III

    (D) It works for all operations

4. (No calculator allowed.) The fraction $\dfrac{4}{9}$ is $\begin{Bmatrix} \text{greater than} \\ \text{equal to} \\ \text{less than} \end{Bmatrix}$ the fraction $\dfrac{1}{2}$.

5. (No calculator allowed.) Fill in the missing numbers in the following sequence by selecting from the choices below. (*Hint*: This is not an arithmetic sequence but rather numbers you should recognize.)

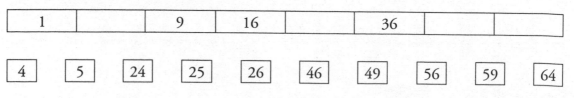

6. A plane leaves Cincinnati and travels at an average speed of 600 miles per hour. After three hours in flight, the plane will have traveled [ ] miles.

7. Simplify: $-3(x + 2x^2 + 3x)$.

    (A) $12x - 6x^2$

    (B) $-3x + 2x^2 - 6x$

    (C) $-18x^2$

    (D) $-12x - 6x^2$

8. The students in the incoming class of 1,250 students at a local college declared their proposed majors from five choices, as shown in the table below.

Major	Percentage
Business	25
Nursing	21
Science	14
Education	10
Liberal Arts	

  a. The percent of students who declared liberal arts is [ ] %.

  b. The number of students who declared business or nursing is [ ].

9. Refer to the graph of the function $F(x) = |x|$ shown below.

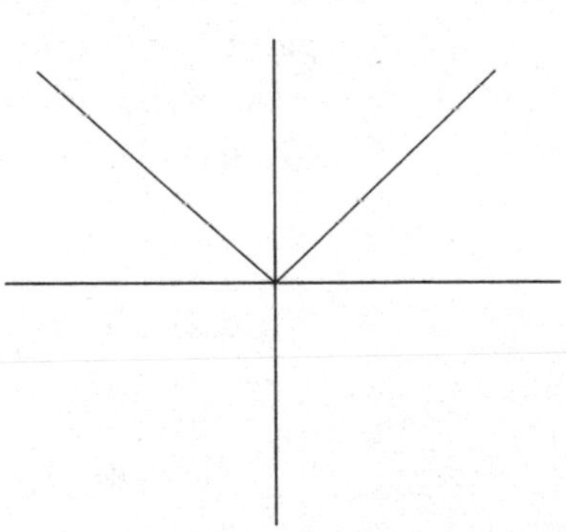

  a. There are $\left\{\begin{array}{c} 0 \\ 1 \\ 2 \end{array}\right\}$ x values for each y value.

  b. There are $\left\{\begin{array}{c} 0 \\ 1 \\ 2 \end{array}\right\}$ y values for each x value.

10. Felicia wants to buy hotdogs for an event to which 150 people are expected. Assume each person will eat two hotdogs. Felicia should buy [＿＿＿＿] hotdogs.

11. A shop owner paid $10.00 for a shirt. He marks it up to make a 70% profit. The price of the shirt is $\left\{ \begin{array}{l} \text{more than} \\ \text{exactly} \\ \text{less than} \end{array} \right\}$ $10.70.

12. For the function $F(x) = 2x^2 - 3x + 5$, what is $F(0)$?

    (A) 0

    (B) 2

    (C) 3

    (D) 5

13. Four gallons of gasoline cost $18.36. The price of gas per gallon is $[＿＿＿＿].

14. A musical performance is sold out in a hall that seats 1,800. If the tickets cost $60 each, and the hall had expenses of $40,000, what was the hall's profit?

    (A) $108,000

    (B) $68,000

    (C) It had a loss of $29,200.

    (D) It broke even.

15. A silo has a cylindrical storage area topped by half of a sphere. To figure out how much paint it will take to cover the whole silo, farmer McGregor needs to know the surface area of the silo. He knows it is 30 feet high and has a radius of 10 feet. What is the surface area (in square feet in terms of π)?

    (A) 600π

    (B) 200π

    (C) 800π

    (D) $1,000\pi$

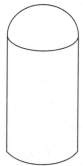

16. Given the five numbers 30, 26, 32, 28, 21, what number can be added to the group so the median of the new set of six numbers is 27?

   (A) 23

   (B) 27

   (C) 29

   (D) 30

17. Find the value of x if $\dfrac{60 \times 7 \times x}{75 \times 14} = 20.$

   (A) 35

   (B) 50

   (C) 3.92

   (D) 78.4

18. We know one angle in triangle JKL is 30°. We also know that one of the other angles is twice the third angle. The sum of the angles in a triangle is 180°. What set of equations will give us the measures of the other two angles?

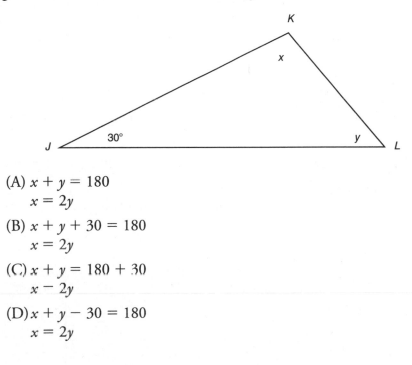

   (A) $x + y = 180$
   $x = 2y$

   (B) $x + y + 30 = 180$
   $x = 2y$

   (C) $x + y = 180 + 30$
   $x - 2y$

   (D) $x + y - 30 = 180$
   $x = 2y$

19. Lin estimated the percentages of his household expenses as shown in the following pie chart.

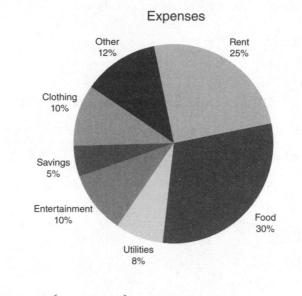

Expenses

a. His food budget is $\left\{\begin{array}{l}\text{equal to}\\ \text{more than}\\ \text{less than}\end{array}\right\}$ what he pays for rent and utillities.

b. If he starts to take $\dfrac{2}{3}$ of the percentage set aside as other to pay off his student loan,

his loan payment would be $\left\{\begin{array}{l}\text{equal to}\\ \text{more than}\\ \text{less than}\end{array}\right\}$ what he budgets for utilities.

20. What is the area of *ABCD* in the figure to the right if each square is one inch on a side?

(A) 12 square inches

(B) 15 square inches

(C) 30 square inches

(D) cannot do this problem with the information given

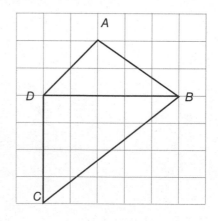

21. Sam buys a used car for $8,000. He puts $500 down and finances the rest for three years at 6% annual interest. At the end of the three years, the $8,000 car will have cost him $ [   ].

22. A regular tetrahedron is actually a pyramid whose sides and base are all equilateral triangles. Each vertex is over the middle of the opposite triangle.

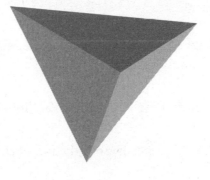

a. A regular tetrahedron has [   ] edges.

b. If an edge is 12 inches and the slant height is $6\sqrt{3}$ inches, what is the surface area of this tetrahedron in square inches?

(A) $144\sqrt{3}$

(B) $288\sqrt{3}$

(C) 144

(D) 288

23. The volume of a cube is 32 cubic units. What is the volume of a smaller cube whose edges are half those of this cube?

(A) 2 cubic units

(B) 4 cubic units

(C) 8 cubic units

(D) 16 cubic units

24. One-third of the students in a GED® prep class has never studied geometry. In a class of 60 students, what is the ratio of those who have studied geometry to those who haven't?

(A) 1:3

(B) 2:1

(C) 1:2

(D) 2:3

25. The hardware store has a sale of 30% off the list price of everything. Dan buys a drill listed at $39.96 and a hammer listed at $9.47. What is the sale price of the drill?

(A) $11.99

(B) $6.63

(C) $27.97

(D) $34.60

26. The world's highest Ferris wheel, called the High Roller, will have 28 passenger cabins, each able to hold 40 people.

a. At full capacity, the High Roller can hold ⬚ people.

b. At a cost of $20 per person, the total ticket sales will be $ ⬚ per ride at full capacity.

27. Write the equation for the following table:

x	1	5	9	13
y	0	2	4	6

(A) $2y = x - 1$

(B) $y = 2x - 2$

(C) $y = \dfrac{1}{2}x + 2$

(D) $2y = x - 4$

28. For which of the following linear equations does a change of 5 units for the x value correspond to a change of 3 units for the y value?

(A) $5y - 3x = 6$

(B) $3y - 5x = 12$

(C) $5y + 3x = 6$

(D) Not enough information is given.

29. If a small spherical steel ball weighs 10 pounds, what does a spherical steel ball twice its diameter weigh?

(A) 20 pounds

(B) 40 pounds

(C) 80 pounds

(D) 100 pounds

30. It took Maureen 12 hours to read a 300-page book. How long will it take her to read a 400-page book if her reading rate is the same?

    (A) 25 hours

    (B) 33 hours and 20 minutes

    (C) 16 hours

    (D) 15 hours and 20 minutes

31. If Janet buys 6 dresses at 15% off the normal price x, which expression represents how much she has saved?

    (A) $15(6x)$

    (B) $0.15(6x)$

    (C) $1.5(6x)$

    (D) $0.15x$

32. Leroy is training for a track meet. Every weekday he runs 6 times around a quarter-mile track. After one week, Leroy has run $\left\{ \begin{array}{l} \text{equal to} \\ \text{more than} \\ \text{less than} \end{array} \right\}$ 12 miles.

33. In an isosceles triangle, the unequal angle is 30 degrees less than either of the other two equal angles. The sum of the angles of a triangle is 180°. What equation would be used to find b, the measure of one of the equal angles?

    (A) $2b + 30 = 180$

    (B) $b + b + b + 30 = 180$

    (C) $b + b + b - 30 = 180$

    (D) $3b = 180$

34. Simplify $8x + 19x - 6y - (-y)$.

    (A) $27x - 5y$

    (B) $27x - 7y$

    (C) $11x - 5y$

    (D) $11x + 7y$

35. Rick drives from Trenton, NJ to Washington, DC, a distance of 177 miles. He then travels on to Richmond, VA, which is 108 miles from Washington. If it takes him 5 hours to complete the entire trip, how fast is he traveling if he is driving at a constant speed?

    (A) 55 mph

    (B) 57 mph

    (C) 62 mph

    (D) 65 mph

36. Six people can sit on a bench in $\left\{ \begin{array}{l} \text{equal to} \\ \text{more than} \\ \text{less than} \end{array} \right\}$ 120 ways.

37. Identify the non-negative whole number values of x that would make $\dfrac{3x+9}{2} < 9$ a true statement by indicating on the following number line.

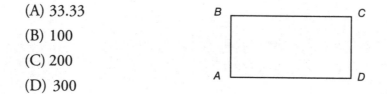

38. You are given rectangle $ABCD$. If the length of this rectangle is tripled and the width stays the same, by what percentage is the area of $ABCD$ increased?

    (A) 33.33

    (B) 100

    (C) 200

    (D) 300

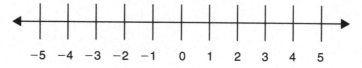

39. The US Americans with Disabilities Act says all public places must be wheelchair accessible. An old restaurant has to retrofit a ramp over two steps to go up to the restrooms. Each step is 7 inches high. The ramp has to have a slope of at most 1 inch of rise to 12

inches of ramp. Therefore, the ramp must be $\left\{ \begin{array}{l} \text{greater than or equal to} \\ \text{less than or equal to} \\ \text{equal to} \end{array} \right\}$ 14 feet.

40. If $\dfrac{2x+6}{2} - x = \dfrac{x^2}{3}$, the value of x is

 (A) ± 3

 (B) ± 9

 (C) $\pm\sqrt{3}$

 (D) 0

41. John rides his bike to work, which is 10 miles away, at an average speed of 15 miles per hour. Jim lives only 2 miles from work, so he walks at a rate of 3 miles per hour, leaving home at the same time as John did. John gets to work $\left\{\begin{array}{c}\text{before}\\\text{after}\\\text{at the same time as}\end{array}\right\}$ Jim.

42. At 11 a.m., water starts to flow into a cylindrical storage tank that is 8 feet in diameter and 12 feet high. If the water is coming into the tank at a rate of 600 cubic feet per hour, at approximately what time will it begin to overflow? (Use $\pi = 3.14$.)

 (A) 11:30 a.m.

 (B) noon

 (C) 1 p.m.

 (D) 3 p.m.

43. A 3×4 rectangle $ABCD$ is plotted on a grid with A at point $(2, 0)$. What is a possible coordinate of point D?

 (A) $(-2, 0)$

 (B) $(5, 0)$

 (C) $(-1, 0)$

 (D) all of these

44. Kurt took a standardized test. Every correct answer was worth 10 points, but every incorrect answer was scored as −5 points (that is, 5 points were subtracted from the final score). Blanks were scored as 0. There were 50 questions on the test. If Kurt answered 30 questions correctly, 10 questions wrong, and left 20 blank, fill in the values in the following equation that calculates Kurt's score. Choose numbers from the choices given and place them in the correct boxes.

| 30 | × | | + | 10 | × | | + | 20 | × | | = | |

| −5 | | 0 | | 5 | | 10 | | 20 | | 30 | | 220 | | 250 | | 270 | | 300 |

45. A car trip is traveled partway on local roads at an average speed of 40 miles an hour and the rest on a highway at an average speed of 70 miles per hour. Three hours of the 400-mile trip is on local roads. The local road distance versus time graph is shown below, where the slope is 40 miles per hour. Use the distance formula ($d = rt$) to figure the highway time once you can calculate the highway distance with information from the local road graph. The trip involved [ ] hours on the highway.

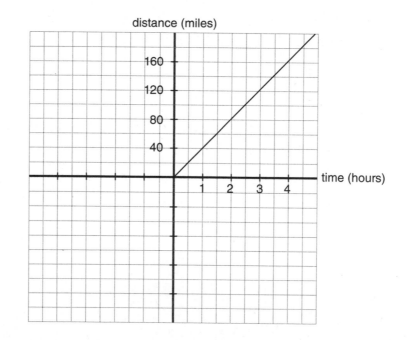

Answer Key – Practice Test 1

1. (B) $8x + 30y$.

2. $0, -2$

3. (B) I and III.

4. Less than.

5. 4, 25, 49, 64.

6. 1800.

7. (D) $-12x - 6x^2$.

8a. 30.

8b. 575.

9a. 2.

9b. 1.

10. 300.

11. More than.

12. (D) 5.

13. $4.59.

14. (B) 68,000.

15. (C) 800π.

16. (A) 23.

17. (B) 50.

18. (B) $x + y + 30 = 180; x = 2y$.

19a. Less than.

19b. Equal to.

20. (B) 15 square inches.

21. $9350.

22a. 6.

22b. (A) $144\sqrt{3}$.

23. 4 cubic units.

24. (B) 2:1.

25. (C) $27.97.

26a. 1,120.

26b. 22,400.

27. (A) $2y = x - 1$.

28. (B) $3y - 5x = 12$.

29. (C) 80 pounds.

30. (C) 16 hours.

31. (B) $0.15(6x)$.

32. Less than.

33. (C) $b + b + b - 30 = 180$.

34. (A) $27x - 5y$.

35. (B) 57 mph.

36. More than.

37. 0, 1, 2.

38. (D) 200.

39. Greater than or equal to.

40. (A) ± 3.

41. At the same time as Jim.

42. (B) Noon.

43. (D) all of these

44. 10, -5, 0, 250.

45. 4 hours.

Detailed Solutions

1. **(B)** $8x + 30y$. Be sure to keep the value with the letter that represents the purchase. His total purchase is the sum of the individual purchases.

2. $0, -2$. Expressions are undefined for any values that make the denominator zero because division by zero is undefined. Here, $x(x + 2) = 0$ when either factor, x or $x + 2$, equals 0.

3. **(B)** I and III. The commutative property does not work for subtraction ($2 - 5 \neq 5 - 2$) nor for division ($4 \div 2 \neq 2 \div 4$), but it does work for multiplication ($2 \times 3 = 3 \times 2$).

4. Less than. There are two ways to look at this problem. One is to recognize that half of the denominator 9 is $4\frac{1}{2}$, so any numerator less than $4\frac{1}{2}$ is smaller. The other way is to convert both fractions to a common denominator, which is 18 in this case. So $\frac{4}{9}$ becomes $\frac{8}{18}$ and $\frac{1}{2}$ becomes $\frac{9}{18}$, and since $8 < 9$, $\frac{4}{9} < \frac{1}{2}$.

5. $4, 25, 49, 64$. This is a sequence of squares: $1^2, 2^2, 3^2, 4^2, 5^2, 6^2, 7^2$.

6. $1,800$. Each hour the plane goes 600 miles, so in 3 hours, it will have traveled $600 \times 3 = 1,800$ miles.

7. **(D)** $-12x - 6x^2$. Remember that bases with unlike exponents cannot be combined.

8a. 30. Add the other categories (total 70%) and subtract from 100%.

8b. 575. The percentage in business or nursing is $25 + 21 = 46$. There were 1,250 total students, so this percentage is $1,250 \times .46 = 575$ students.

9a. 2. There are two x values for each y value, except at $(0, 0)$.

9b. 1. Because this is a function, there is only one point in the range for each point in the domain.

10. 300. If 150 people each eat 2 hotdogs, the total is $150 \times 2 = 300$ hotdogs.

11. **More than.** A 70% profit on $10.00 is $10 × .70 = $7. He must add the profit to the cost of the shirt, so he prices the shirt at $10 + $7 = $17.00.

12. **(D) 5.** If $x = 0$ in $2x^2 - 3x + 5$, the first two terms become 0 and only the 5 is left.

13. **$4.59.** For dollars per gallon, divide gallons into dollars.

14. **(B) 68,000.** The income from the tickets is ($1,800 × 60) = $108,000. The profit is income minus expenses, $108,000 − $40,000 = $68,000.

15. **(C) 800π.** The surface area of the silo is only the portion that can be seen. So it is the area of the cylinder without the circular ends plus the surface area of half a sphere. The cylindrical area is actually a rectangle that is 30 feet high and has a width equal to the circumference of the silo. The surface area of the cylindrical part is $SA_c = 2\pi rh = 2\pi(10)(30) = 600\pi$. The surface area of the dome is $SA_d = (\frac{1}{2})(4\pi r^2) = 2\pi(10)^2 = 200\pi$. So the surface area of the silo is $600\pi + 200\pi = 800\pi$.

16. **(A) 23.** When the numbers are written in order, the median of the original set of numbers is 28. For the new median to be 27, a number has to be added to the lower end of the set of ordered numbers. That eliminates answer choices (C) and (D). If answer choice (B), 27, is added, the new median will be the average of the middle two numbers, 27 and 28, which is 27.5, not 27. If answer choice (A), 23, is added, the new median will be the average of the middle two numbers, 26 and 28, which is 27.

17. **(B) 50.** This problem can be done on the calculator, but you must remember to put parentheses around the numbers or you will end up with answer choices (C) or (D), which are wrong. The calculator entry for the fraction should look like (60 × 7) ÷ (75 × 14) = 0.4, and the equation becomes $0.4x = 20$, so $x = \frac{20}{.4} = 50$. Another way to do the problem is to cancel common factors of 7, 15, and 2 from the left-hand side of the equation, which yields $\frac{2}{5}x = 20$, so $x = 50$.

18. **(B)** $x + y + 30 = 180; x = 2y$. The second equations in all of the answer choices are the same, and they are correct for "one of the other angles is twice the third angle." So look at the first equations. The angles in a triangle sum to 180°, so the first equation must say that $x + y + 30 = 180$. Answer choice (B) is the only one that does. The other answer choices may have the correct terms but not necessarily the correct signs for them.

19a. **Less than.** The percentage for rent is 25% and for utilities is 8%, totaling 33% of Lin's budget. Food is 30%, so it is less than the sum of the other two.

19b. Equal to. The percentage set aside as other is 12%, and $\frac{2}{3}$ of that is $\frac{2}{3} \times 12\% = 8\%$. Lin budgets 8% for utilities, so the two percentages are equal.

20. **(B)** 15 square inches. The figure is a composite of two triangles, both with base DB. The formula for the area of a triangle is $A = \frac{1}{2}bh$. This formula is not on the GED® test formula sheet but is one you should know. The triangle on the top has an area of $A = \frac{1}{2}(5)(2) = 5$ square inches, and the one on the bottom has an area of $A = \frac{1}{2}(5)(4) = 10$ square inches. So the total area of $ABCD$ is $5 + 10 = 15$ square inches. Since this figure is drawn to scale, you also could have just counted the squares, but be careful with estimating all those partial squares.

21. $9,350. Notice that he financed $7,500 because he paid $500 as a down payment. Use $I = prt = (7,500)(.06)(3) = \$1,350$. So the car cost him $8,000 + \$1,350 = \$9,350$ after three years. Be sure to fill in only the numbers because the dollar sign is provided.

22a. 6. There are three edges for the base and three edges that go to the vertex.

22b. **(A)** $144\sqrt{3}$. The formula for the surface area of a right pyramid is $SA = \frac{1}{2}ps + B$, which is on the GED® test formula sheet. The slant height is s, p is the perimeter of the base, and B is the area of the base. Here, the slant height is given as $s = 6\sqrt{3}$ inches, p is $3(12) = 36$ inches, and the base is an equilateral triangle with base 12 and height $6\sqrt{3}$, so the area of the base $B = \frac{1}{2}(12)(6\sqrt{3})$ $= 36\sqrt{3}$. Therefore, $SA = \frac{1}{2}(36)(6\sqrt{3}) + 36\sqrt{3} = 108\sqrt{3} + 36\sqrt{3} = 144\sqrt{3}$. This problem could have been done more quickly by recognizing that the surface area is just four times the area of one of the faces because all the faces are equilateral triangles with base 12 and height $6\sqrt{3}$. So the surface area is $4(\frac{1}{2})(12)(6\sqrt{3}) = 144\sqrt{3}$.

23. 4 cubic units. The ratio of the edges of the two cubes is given as 2:1. The ratio of their volumes is then $(2)^3 : (1)^3$, or 8:1. So the proportion for their volumes is $\frac{8}{1} = \frac{32}{x}$, which by cross-multiplication gives $8x = 32$, or $x = 4$, the volume of the smaller cube. You don't have to find the lengths of the edges, just the ratios of the volumes.

24. **(B)** 2:1. The problem gives the information that one-third of the students had not studied geometry, so that would be $\frac{1}{3} \times 60 = 20$ students who had not studied geometry, leaving 40 who have. It asks for the ratio of those who have to those who have not, so the ratio is 40:20 = 2:1. Another way to look at this problem is that if $\frac{1}{3}$ have not, then $\frac{2}{3}$ have, and the ratio is $\frac{2}{3} : \frac{1}{3} = 2{:}1$ as well.

25. **(C) $27.97.** Read this carefully. You are given the price of a drill and a hammer, but you are asked only about the drill—this problem doesn't have anything to do with the hammer, so answer choices (A) and (D) are eliminated. A 30% discount on a $39.96 drill is $11.99, but you are asked the sale price, which is $39.96 − $11.99 = $27.97.

26a. **1,120.** At full capacity the High Roller will hold $28 \times 40 = 1{,}120$ people.

26b. **22,400.** If each pays $20, the income will be $20 \times 1{,}120 = $22{,}400$. The dollar sign is already in the question, so enter only the number of dollars.

27. **(A) $2y = x − 1$.** For every change of 2 in y, there is a change of 4 in x, so the slope is $m = \dfrac{\text{change in } y}{\text{change in } x} = \dfrac{2}{4} = \dfrac{1}{2}$. Answer choice (A) is correct because, in slope-intercept form, it is $y = \dfrac{x}{2} - \dfrac{1}{2}$, so the slope is correct. The y-intercept is also correct because, according to the pattern in the table, when $x = 0$, y would be $-\dfrac{1}{2}$. Answer choice (B) is eliminated because it has the wrong slope, answer choice (C) is eliminated because the y-intercept must be negative according to the pattern in the table, and answer choice (D) is eliminated because when $y = -2$, x would be 0, and that also doesn't fit the pattern in the table.

28. **(B) $3y − 5x = 12$.** A change of 5 units for the x value corresponding to a change of 3 units for the y value is the same as saying the slope of the line is $\dfrac{5}{3}$, so putting each equation into slope-intercept form shows that $3y - 5x = 12$ is the same as $y = \dfrac{5}{3}x + 4$, which has a slope of $\dfrac{5}{3}$.

29. **(C) 80 pounds.** The ball is twice the size of the 10-pound ball, but its weight is 8 times as much. This is because the radius is doubled, and that is raised to the third power. The weight of the ball is related to its volume. The formula for the volume of a sphere (on the GED® formula sheet) is $V = \dfrac{4}{3}\pi r^3$, and if we double the size, r becomes $2r$. The new volume is $V_{\text{new}} = \dfrac{4}{3}\pi(2r)^3 = 8V_{\text{old}}$.

30. **(C) 16 hours.** There are several ways to solve this problem. One is to set up a proportion: if it took 12 hours to read 300 pages, it will take x hours to read 400 pages, or $\dfrac{12}{300} = \dfrac{x}{400}$, so

$x = 16$. Another way is to find the rate of pages per hour ($\frac{300}{12} = 25$) and divide that into 400:

$\frac{400}{25} = 16$.

31. **(B)** 0.15(6x). The decimal 0.15 is the same as 15%, and $6x$ represents the normal price of 6 dresses.

32. Less than. Every weekday he runs 6 times around a quarter-mile track. Read this carefully. In one week, Bill is running 5 days, not 7. The calculation is $5 \times 6 \times .25 = 7.5$ miles, which is less than 12 miles.

33. **(C)** $b + b + b - 30 = 180$. The equal angles are each b degrees, and the third angle is 30 degrees less than either one of them, so its measure is $(b - 30)$. Use an equation that says the three angles in a triangle must total 180 degrees, so $b + b + (b - 30) = b + b + b - 30 = 180$.

34. **(A)** $27x - 5y$. Remember that subtracting a negative number makes it a positive.

35. **(B)** 57 mph. The distance formula is $d = rt$. To find the rate, divide the distance by the time. The distance is the total for the two trips, which is $177 + 108 = 285$ miles. Then the rate is $285 \div 5 = 57$ mph.

36. More than. The first person on the bench can be picked in 6 ways. For the next position, there are only 5 people left, so it is 5 ways, and so forth. The calculation is $6 \times 5 \times 4 \times 3 \times 2 \times 1 = 720$. If you recognized that this is a permutation of 6 people, with a value of 6! (factorial), which is more than 5! = 120, all the better.

37. 0, 1, 2. Multiply through by 3 to get rid of the fraction: $3x + 9 < 18$, then $3x < 9$, and $x < 3$. But do not include any negative numbers because the question asks for non-negative numbers only.

38. **(D)** 200. The original area is $A = lw$. The new area is $A' = (3l)w = 3lw = 3A$. In other words, it is tripled, which means increased by 200%. You might think it is 300%, but the question isn't asking what the new area is, only by what percentage it is *increased*. The increase from A to $3A$ is $2A$, which is 200%. Read carefully.

39. Greater than or equal to. The ratio is 12 inches of ramp for each rise of 1 inch, which is the same as 1 foot per 1 inch. Since the rise is 2(7 inches) for the two steps, 14 inches of rise requires 14 feet of ramp. A ramp any shorter will have too large a slope.

40. **(A)** ± 3. Reduce the first term by canceling the 2's to get $x + 3$ as the first term. Multiply both sides by 3 to clear the equation of fractions, $3(x + 3 - x) = x^2$. Then combine terms, $9 = x^2$. Finally, put everything on the same side of the equation as the x^2 term, with 0 on the other side, and this is the difference of two squares, $0 = x^2 - 9$, so $x = \pm 3$.

41. At the same time as Jim. It takes John $\dfrac{10}{15} = \dfrac{2}{3}$ of an hour to get to work. It takes Jim $\dfrac{2}{3}$ of an hour to get to work. They both arrive 40 minutes after they leave home.

42. **(B)** Noon. The tank holds a capacity (volume) of $V_{\text{cylinder}} = \pi r^2 h$ cubic feet of water. That equation is on the GED® test formula sheet. The tank will start to overflow when that much water has been pumped in. Remember that you need the radius, not the diameter, for this formula. So the capacity of the tank is $V = (3.14)(4)^2(12) = 602.88$ cubic feet (use the calculator). Therefore, since the water is coming in at 600 cubic feet per hour, the tank will start to overflow approximately one hour after 11 a.m., which is noon.

43. **(D)** All of these. If A is point (2, 0) on a 3 × 4 rectangle, D can be any of four points ± 3 or ± 4 units from A. They will all have the same y value as A, which is 0. So the four choices are (-2, 0), (-1, 0), (5, 0), or (6, 0). Three of these are listed as answer choices (A), (B), and (C), so the answer is all of these.

44. 10, -5, 0, 250 (in that order). To calculate Kurt's score, multiply how many of each type of answer (correct, wrong, blank) by its value and add them, so 30 (correct answers) × 10 (points) would give the score for the correct answers; 10 (wrong answers) × -5 (points) for the wrong answers; and 20 (blanks) × 0 (points) for the wrong answers, or

$$30 \times 10 + \underline{20} \times \underline{-5} + 20 \times \underline{0} = \underline{250}$$

Note that although parentheses would be helpful in this equation, they are not necessary because the order of operations says to do multiplications before additions.

45. 4 hours. The graph of local road travel has a slope of 40 miles per hour. After three hours, the distance is 120 miles. That means 280 miles are driven on the highway at a rate of 70 miles per hour. Therefore, using $d = rt$, we get $280 = 70t$, so $t = 4$ hours are driven on the highway.

Answer Sheet – Practice Test 1

1. Ⓐ Ⓑ Ⓒ Ⓓ

2. _____

3. Ⓐ Ⓑ Ⓒ Ⓓ

4. _____

5. _____

6. [_____]

7. Ⓐ Ⓑ Ⓒ Ⓓ

8a. [_____]

8b. [_____]

9a. _____

9b. _____

10. [_____]

11. _____

12. Ⓐ Ⓑ Ⓒ Ⓓ

13. [_____]

14. Ⓐ Ⓑ Ⓒ Ⓓ

15. Ⓐ Ⓑ Ⓒ Ⓓ

16. Ⓐ Ⓑ Ⓒ Ⓓ

17. Ⓐ Ⓑ Ⓒ Ⓓ

18. Ⓐ Ⓑ Ⓒ Ⓓ

19a. _____

19b. _____

20. Ⓐ Ⓑ Ⓒ Ⓓ

21. [_____]

22a. [_____]

22b. Ⓐ Ⓑ Ⓒ Ⓓ

23. Ⓐ Ⓑ Ⓒ Ⓓ

24. Ⓐ Ⓑ Ⓒ Ⓓ

25. Ⓐ Ⓑ Ⓒ Ⓓ

26a. [_____]

26b. [_____]

27. Ⓐ Ⓑ Ⓒ Ⓓ

28. Ⓐ Ⓑ Ⓒ Ⓓ

29. Ⓐ Ⓑ Ⓒ Ⓓ

30. Ⓐ Ⓑ Ⓒ Ⓓ

31. Ⓐ Ⓑ Ⓒ Ⓓ

32. _____

33. Ⓐ Ⓑ Ⓒ Ⓓ

34. Ⓐ Ⓑ Ⓒ Ⓓ

35. Ⓐ Ⓑ Ⓒ Ⓓ

36. _____

37. _____

38. Ⓐ Ⓑ Ⓒ Ⓓ

39. _____

40. Ⓐ Ⓑ Ⓒ Ⓓ

41. _____

42. Ⓐ Ⓑ Ⓒ Ⓓ

43. Ⓐ Ⓑ Ⓒ Ⓓ

44. _____

45. [_____]

GED® MATH REASONING

Practice Test 2

The GED® Math Reasoning test has five types of questions: multiple-choice, drop-down, fill-in-the-blank, hot-spot, and drag-and-drop. We represent all these question types in our practice tests. Please refer to the question-type descriptions starting on page 4 to refresh your memory of the question types on this test.

1. (No calculator allowed.) Another word for *mean* is

   (A) median.
   (B) mode.
   (C) average.
   (D) range.

2. (No calculator allowed.) Which of these numbers is not in scientific notation?

   (A) 2.34×10^{-34}
   (B) -2.34×10^{-34}
   (C) $5.67 \times 10^{3.3}$
   (D) 8.43×10^{10000}

3. Fill in the missing numbers in the correct places in the following geometric sequence by selecting numbers from the choices given.

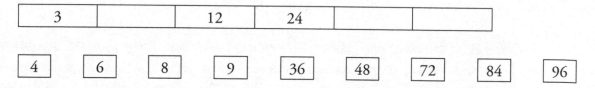

| 3 | | 12 | 24 | | |

| 4 | 6 | 8 | 9 | 36 | 48 | 72 | 84 | 96 |

4. (No calculator allowed.) The percentage equivalent of the decimal .964 to the nearest tenth is ⬚ %.

5. (No calculator allowed.) The median of the following data is:

   2   3   4   5   6   7   8   9   10   10   10   15   15   15   35   45

   (A) 10
   (B) 9
   (C) 9.5
   (D) 10 and 15; it is bimodal

6. Simplify by combining terms: $(2x^3 - 3x^2 + x) - (x^2 - 1) + 4x^2$.

   (A) $2x^3 + x + 1$
   (B) $2x^3 + x - 1$
   (C) $2x^3 + x^2 + x + 1$
   (D) $2x^3 - x + 1$

7. When Suzanne was traveling on Route 95 she saw a sign for her exit indicating that it was $\frac{1}{4}$ mile ahead. Actually, it was only $\frac{1}{8}$ mile ahead. The difference between what the sign had posted and the actual distance is [＿＿＿＿] feet. (One mile is 5,280 feet.)

8. If the diameter of circle O is 10 inches, the area (in terms of π) is [＿＿＿＿] π square inches.

9. The results of a random survey of voters with regard to legalizing marijuana are shown in the table below.

Age Group	In Favor	Opposed	No Opinion
21–30	33	15	2
31–40	32	12	6
41–50	28	24	3
51–60	30	23	2
61–70	7	14	3
Over 70	11	23	4

a. In the 21–40 age groups, what percentage supports legalizing marijuana?

(A) 12%

(B) 32%

(C) 50%

(D) 65%

b. Expressed as a fraction in lowest terms, the ratio of those in favor to those opposed in the 41–50 age group is [＿＿＿＿]. (Use a slash for the fraction bar.)

c. In the 61–70 age group, the number opposed is $\left\{ \begin{array}{c} \text{less than} \\ \text{half} \\ \text{exactly twice} \end{array} \right\}$ the number in favor.

10. Clarissa wants to prop a large picture, measuring 6 feet high and 4 feet wide, against the wall instead of hanging it. She has to prop it very close to the wall so it won't slip on the carpet. If she thinks that the bottom of the picture should be a foot from the wall, how high on the wall will the top of the picture rest?

(A) 5.8 feet

(B) 5.9 feet

(C) 6 feet

(D) 6.1 feet

11. How many gallon cans of paint should Connor buy to cover the four walls of a room that is 9 feet high by 12 feet wide by 15 feet long if the label says one gallon will cover 350 square feet? The paint is sold only in one-gallon containers.

    (A) 1

    (B) 2

    (C) 3

    (D) 4

12. A concert is held in a stadium that seats 17,250 people. If the stadium sold 88% of the available tickets, [   ] tickets were sold and [   ] tickets were unsold.

13. If a Ferris wheel takes 30 minutes to make one complete turn, how many minutes after boarding would a passenger be halfway to the top?

    (A) 7.5 minutes

    (B) 15 minutes

    (C) 30 minutes

    (D) 45 minutes

14. An advertiser for a mining company wants to cover a sphere with gold leaf. His budget allows for a sheet of gold leaf that is only 10 inches on a side. What is the radius of the largest sphere the advertiser can afford? (Use $\pi = 3.14$, and round the answer to the nearest tenth.)

    (A) 5.6 inches

    (B) 3.2 inches

    (C) 2.8 inches

    (D) 1.4 inches

15. To produce $450 in interest, $3,000 should be invested at 5% interest for [   ] years.

16. Trevor got a large rectangular aquarium for his pet tortoise. The aquarium measures 30 inches long, 20 inches wide, and 20 inches high. He wants to cover the bottom of the aquarium with sand to a depth of 6 inches. How many 1-cubic-foot bags of sand will give him an amount of sand that is closest to his 6-inch depth?

    (A) 1

    (B) 2

    (C) 3

    (D) 4

17. Boxplots show

    (A) the median.

    (B) the first and third quartiles.

    (C) the minimum and maximum data points.

    (D) all of these.

18. Scott is taking a bike trip. He wants to go 125 miles in 7.5 hours. Indicate by a dot on the number line below Scott's average speed.

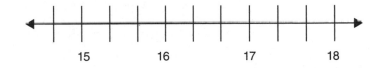

19. Combine: $\dfrac{4(x-1)+4(-3x+5)}{2x-2(x+4)} =$

    (A) $2 - x$

    (B) $x - 2$

    (C) $2x - 2$

    (D) $-2x + 2$

20. The value of $2x^2 - 3xy - 4y^2$ when $x = -3$ and $y = 5$ is $\boxed{}$.

21. The volume of a cube is 64 cubic inches. The edge of a smaller cube is 1 inch. What is the ratio of the areas of the faces of these two cubes?

    (A) 2:1

    (B) 4:1

    (C) 8:1

    (D) 16:1

22. Which of the following is the same as $-|-9-(-5)|$?

    (A) $|-9-5|$

    (B) $|9+5|$

    (C) $-|9+5|$

    (D) $-|-9+5|$

23. A rate of 60 miles per hour is $\left\{ \begin{array}{l} \text{faster than} \\ \text{the same as} \\ \text{slower than} \end{array} \right\}$ a mile a minute.

24. The batting average for a baseball player is expressed as the decimal equivalent of the percentage of base hits to the number of times at bat (not counting walks, sacrifices, or being hit by the ball).

   a. The scoreboard indicates that the third batter's average was .250. After his next time at bat, his batting average changed to .235. He probably $\left\{ \begin{array}{l} \text{made a hit} \\ \text{struck out} \\ \text{walked} \end{array} \right\}$.

   b. Glen has been at bat 55 times this season, and his batting average is .291. He has made [   ] base hits.

   c. Shawn has been at bat 100 times this season, and his batting average is .250. How many times has he walked?

      (A) 25
      (B) 75
      (C) 0
      (D) cannot tell

25. The number of women who work as flight attendants for a regional airline is 10 less than twice the number of men. If there are 140 flight attendants employed by the airline, how many are men?

      (A) 40
      (B) 50
      (C) 90
      (D) 100

26. Nancy wants to hang a mirror centered between two windows. The mirror will be held in place by clips at each of its corners. The size of the mirror is 2 feet by 5 feet, and the space between the windows is 42 inches. How far from the windows will the clips be placed?

    (A) 9 inches

    (B) 14 inches

    (C) 18 inches

    (D) 28 inches

27. Given rectangle *EFGH* with point *E* at (1, 6) and point *G* at (5, 3). The scale of both grids is 1 unit. The slope of diagonal *EG* is

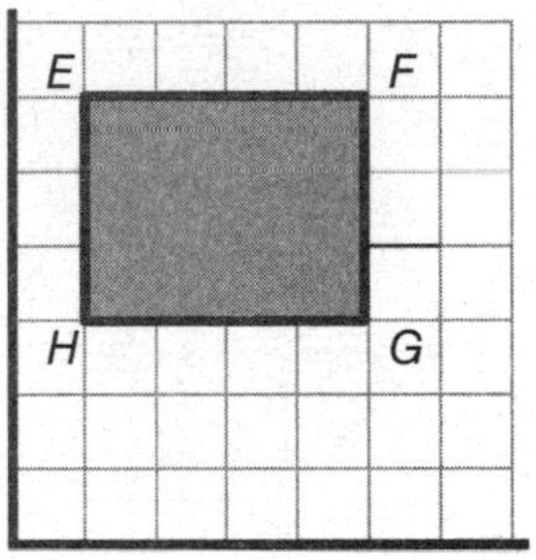

    (A) $\dfrac{3}{4}$

    (B) $-\dfrac{3}{4}$

    (C) $\dfrac{4}{3}$

    (D) $-\dfrac{4}{3}$

28. The function $F(x)$ has the following features: x-intercept at $(3, 0)$; y-intercept at $(0, 2)$; endpoint at $(-4, 4)$; increasing in the interval from $x = -4$ to $x = -2$ with a maximum at $(-2, 6)$, then a negative slope to $(0, 2)$; and symmetry in the interval $(3, 0)$ to $(9, 0)$ around the line $x = 6$. Which graph is the graph of this composite function?

(A) I

(B) II

(C) III

(D) IV

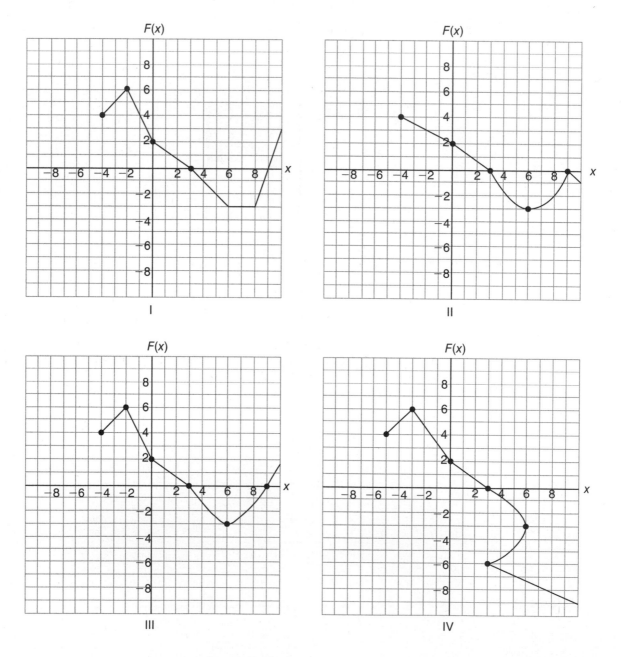

29. Simplify: $7x^3 + 5y^2 - 4y^2 + 2x$.

    (A) $9x^4 + 1$

    (B) $9x^4 + y^4$

    (C) $9x^2 + y^2$

    (D) $7x^3 + y^2 + 2x$

30. All of the steps in the figure below are 2 units high with 2-unit risers. All the angles are right angles. What is the area of the figure?

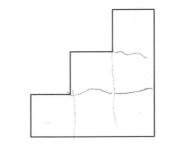

    (A) 18 square units

    (B) 22 square units

    (C) 24 square units

    (D) 36 square units

31. Amber is in high school and wants to save $1,800 toward her future college expenses. If Amber has a part-time job that pays $9.00 per hour, she works 20 hours per week, and she saves all of her earnings, it will take her $\left\{\begin{array}{l}\text{less than}\\\text{more than}\\\text{exactly}\end{array}\right\}$ three months to reach her goal.

32. Indicate by dot(s) the value(s) on the number line for which all of the following expressions is undefined:

$$\frac{x-3}{x^2-9} \qquad (x-3)^0 \qquad \frac{x^2-3}{9-x^2}$$

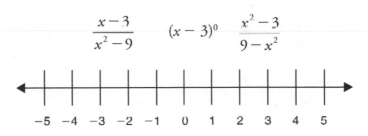

33. A rectangle has dimensions of 4 inches by $2\frac{1}{2}$ inches. If one of the diagonals is drawn, the area of the triangle formed by two sides of the rectangle and the diagonal is [ ] square inches.

34. The numerical expression $\dfrac{2^3 + 4^2 - 2^2 - (4-2)^2}{2}$ is equivalent to

    (A) 4

    (B) 8

    (C) 14

    (D) 16

35. Five men and five women put their names on slips of paper, and three slips are drawn at random for an elite committee. One of the men is named Jeff. The probability that Jeff's name is not selected at all is [ ]. Give your answer as a fraction, using a slash for the fraction line.

36. Which of the following equations, together with the equation $y = 2x - 3$, is sufficient for finding the values of x and y?

    (A) $y = 2x + 5$

    (B) $2y = 4x$

    (C) $y = 3 + 2x$

    (D) $y = 3x - 2$

37. Felix scored 6 more points than Billy in a basketball game, and their scores together made up a half of the team's total scoring, which was 80 points. Which of the following equations would find how many points Billy scored?

    (A) $(6 + x) = \dfrac{80}{2}$

    (B) $(2x + 6) = \dfrac{80}{2}$

    (C) $\dfrac{80}{2} = 6 - x$

    (D) $x = \dfrac{80}{2 \times 6}$

38. Ten slips of paper numbered 1 to 10 are placed in a box. The probability of selecting an odd number and then an even number with no replacement is $\left\{ \begin{array}{l} \text{greater than} \\ \text{less than} \\ \text{the same as} \end{array} \right\}$ the probability of selecting an odd number and then an even number with replacement.

39. Five people want to sit on a bench. One person absolutely wants to sit on either end of the bench. In how many ways can the five people sit on the bench?

    (A) 5

    (B) 12

    (C) 24

    (D) 48

40. Fill in the blanks by selecting from the given choices to make the following equation true for $x = 4$.

(	x		)	(	x		)	=	8

+4	−3	−2	−1	+1	+2	+3	+4

41. The length of a rectangle is four times the width. The area is 16 square units. The width of the rectangle is [ ] units.

42. Consider the function $F(x) = x^2 - 3x + 7$. Arrange the following values of this function in order from largest to smallest: $F(3)$, $F(1)$, $F(-2)$, $F(-1)$.

    (A) $F(-2) > F(1) > F(3) > F(-1)$.

    (B) $F(-2) > F(-1) > F(3) > F(1)$.

    (C) $F(-2) > F(3) > F(1) > F(-1)$.

    (D) $F(-2) > F(3) > F(-1) > F(1)$.

43. A gallon of gas costs $3.90. Jason's car holds 15 gallons and gets 20 miles to the gallon.

    a. Jason can travel [ ] miles on one tank of gas.

    b. If Jason drives from Phoenix to Denver (950 miles), he will have to stop for gas [ ] times, assuming he starts with a full tank.

44. In square *MNOP*, the sides are of length 4.

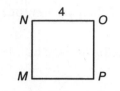

a. If the coordinates of *M* are (1, 1), the coordinates of *O* are

(A) (1, 5)

(B) (5, 5)

(C) (5, 1)

(D) (4, 4)

b. The slope of *MO* is [   ], and the slope of *NP* is [   ].

45. What are the real roots of the quadratic equation $x^2 + 9 = 0$?

(A) ±3

(B) +3

(C) −3

(D) It has no real roots.

1. (C) average.

2. (C) $5.67 \times 10^{3.3}$.

3. 6, 48, 96.

4. 96.4.

5. (C) 9.5.

6. (A) $2x^3 + x + 1$.

7. 660.

8. 25.

9a. (D) 65%.

9b. 7/6.

9c. Exactly twice.

10. (B) 5.9 feet.

11. (B) 2.

12. 15,180; 2,070.

13. (A) 7.5 minutes.

14. (C) 2.8 inches.

15. 3 years.

16. (B) 2.

17. (D) all of these.

18. $16\frac{2}{3}$ (the mark just before the 17 on the number line).

19. (B) $x - 2$.

20. -37.

21. (D) 16:1.

22. (D) $-|-9 + 5|$.

23. The same as.

24a. Struck out.

24b. 16.

24c. (D) cannot tell.

25. (B) 50.

26. (A) 9 inches.

27. (B) $-\dfrac{3}{4}$.

28. (C) III.

29. (D) $7x^3 + y^2 + 2x$.

30. (C) 24 square units.

31. Less than.

32. 3.

33. 5.

34. (B) 8.

35. 7/10.

36. (D) $y = 3x - 2$.

37. (B) $(2x + 6) = \dfrac{80}{2}$.

38. Greater than.

39. (D) 48.

40. $-3, 4$.

41. 2 units.

42. (B) $F(-2) > F(-1) > F(3) > F(1)$.

43a. 300.

43b. Three.

44a. (B) (5, 5).

44b. $1; -1$.

45. (D) It has no real roots.

Detailed Solutions

1. **(C)** Average. The mean is computed the same as an average: $\dfrac{\text{sum of data points}}{\text{number of data points}}$.

2. **(C)** $5.67 \times 10^{3.3}$. In scientific notation, the exponent of 10 has to be a whole number. The leading decimal can be negative, so (B) is okay.

3. 6, 48, 96. This sequence is obtained by doubling each succeeding number.

4. 96.4. To change a decimal to a percentage, move the decimal point to the right. Do not put a % sign in the answer box because it is already provided.

5. **(C)** 9.5. There are 16 data points, so the median will be the average of the 8th and 9th points, which are 9 and 10. Their average is 9.5. Answer choice (D) would be true if the question was about the mode, not the median.

6. **(A)** $2x^3 + x + 1$.

   $$(2x^3 - 3x^2 + x) - (x^2 - 1) + 4x^2$$

   Remove parentheses: $\qquad 2x^3 - 3x^2 + x - x^2 + 1 + 4x^2$

   Combine like terms: $\qquad 2x^3 + x + 1$

7. 660. The difference is $\dfrac{1}{4} - \dfrac{1}{8} = \dfrac{1}{8}$ mile. In feet, that is $\dfrac{1}{8} \times 5{,}280 = 660$.

8. 25. If the diameter of a circle is 10 inches, the radius (r) is half that, or 5 inches, and the area is given by $A = \pi r^2$. So $A = 5^2 \pi = 25\pi$. Do not write π square inches because that is already written in the problem.

9a. **(D)** 65%. There are 100 people in the $21-40$ age groups, and of these, 65 support legalizing marijuana, so the percentage is $65 \div 100 = 65\%$.

9b. 7/6. In the $41-50$ age group, there are 28 in favor and 24 opposed. The ratio in fractional form is $\dfrac{28}{24}$. Both numerator and denominator are divisible by 4, so this reduces to $\dfrac{7}{6}$.

9c. Exactly twice. The number opposed is 14 and the number in favor is 7, so the number opposed is exactly twice the number in favor in the 61−70 age group.

10. **(B)** 5.9 feet. The picture, the wall, and the carpet will form a right triangle, so use the Pythagorean Theorem to solve this problem. The width of the picture isn't used here, just the length, which is the hypotenuse of the right triangle. The wall and floor form a right angle, and the length of the side of the triangle that is the distance of the picture from the wall is 1 foot. So $c = 6$, $a = 1$, and the Pythagorean Theorem gives $(6)^2 = (1)^2 + b^2$, or $b^2 = 35$, $b = \sqrt{35}$, which can be found on the calculator to be 5.92, which rounds to 5.9 feet.

11. **(B)** 2. Two of the walls are (12×9) square feet and two are (15×9) square feet. So the total area to be painted is $2(12 \times 9) + 2(15 \times 9) = 2(108) + 2(135) = 2(243) = 486$ square feet. Connor has to buy 2 cans of paint.

12. 15,180; 2,070. This is a straightforward calculation: 88% is .88, and $(.88)(17,250) = 15,180$. To calculate the number of unsold tickets, subtract 15,180 from 17,250 to get $17,250 − 15,180 = 2,070$. Another way would be to subtract 88% from 100%, to get 12% = .12, and $(.12)(17,250) = 2,070$.

13. **(A)** 7.5 minutes. Halfway to the top means one-quarter of the way around the circle, so it would take one-quarter of the time. If it takes 30 minutes for a complete turn, then a passenger would be halfway to the top at $(.25)(30) = 7.5$ minutes. Answer choice (B) is the time to get to the top.

14. **(C)** 2.8 inches. The surface area of the sphere has to be equal to or less than $10^2 = 100$ square inches. The formula for the surface area of a sphere (from the GED® test formula sheet) is $SA = 4\pi r^2$. Therefore, $4\pi r^2 \leq 100$, or $r^2 \leq \dfrac{25}{\pi}$. Thus, $r \leq \sqrt{7.96} \leq 2.8$ inches.

15. 3 years. $3,000 invested at 5% yields interest of $3,000 \times .05 = $150 per year. So in three years it will yield $450.

16. **(B)** 2. First, figure how many cubic inches of sand Trevor needs to cover the aquarium to a 6-inch depth. That is $30 \times 20 \times 6 = 3,600$ cubic inches. Each bag of sand is $12^3 = 1,728$ cubic inches because 1 foot = 12 inches. Two bags will thus be $2(1,728) = 3,456$ cubic inches, the closest to his desired 3,600 cubic inches.

17. **(D)** All of these. Boxplots are a convenient way to visualize all of the above aspects of a set of data plus any outliers.

18. $16\frac{2}{3}$ (the mark just before the 17 on the number line).

19. **(B)** $x - 2$. Remove the parentheses first.

$$\frac{4(x-1)+4(-3x+5)}{2x-2(x+4)} =$$

$$\frac{4x-4-12x+20}{2x-2x-8} =$$

$$\frac{-8x+16}{-8} = x-2$$

20. -37. Substituting in the values of x and y gives $2(-3)^2 - (3)(-3)(5) - 4(5)^2 = 2(9) - (-45) - 4(25) = 18 + 45 - 100 = 63 - 100 = -37$.

21. **(D)** 16:1. The edge of the larger cube is $\sqrt[3]{64} = 4$. So the ratio of the edges of the cubes is 4:1, and the ratio of the area of their faces is then $4^2 : 1^2$, or 16:1.

22. **(D)** $-|-9 + 5|$. You don't have to actually get a value for the expression. The expression will be a negative since it is an absolute value (positive) preceded by a negative sign. So you can eliminate answer choices (A) and (B) right away because they are positive answers. Now look at the values inside the absolute value signs in (C) and (D) to see which matches the given expression.

23. The same as. The ratio is $\dfrac{60 \text{ miles}}{1 \text{ hour}} = \dfrac{60 \text{ miles}}{60 \text{ minutes}}$, which is the equivalent of a mile a minute.

24a. Struck out. The batting average went down, so he couldn't have made a hit, or it would have gone up. If he walked, the batting average would stay the same because walks aren't calculated in batting averages.

24b. 16. The batting average is calculated as batting average $= \dfrac{\text{base hits}}{\text{times at bat}}$. For Glen, the numbers are $.291 = \dfrac{\text{base hits}}{55}$, or base hits $= .291(55) = 16$.

24c. **(D)** Cannot tell. As stated, walks are not included in the batting average.

25. **(B)** 50. The problem can be restated as "the number of women and men equals 140." Algebraically, this is $(2m - 10) + m = 140$, which reduces to $3m = 150$, so $m = 50$ men.

26. **(A)** 9 inches. The space between the windows is given as 42 inches. The width of the mirror is 2 feet = 24 inches. So the space left is $42 - 24 = 18$ inches, which allows 9 inches from the window on each side of the mirror.

27. **(B)** $-\dfrac{3}{4}$. Slope is $\dfrac{\text{rise}}{\text{run}}$. From point E to point G, this is $\dfrac{-3}{4} = -\dfrac{3}{4}$.

28. **(C)** III. Graph III is the only one that has all of the features that are described. Eliminate graph IV right away because it isn't a function (there are two y values for $3 > x > 6$). Graph I lacks the symmetry around $x = 6$, and graph II lacks the maximum at $(-2, 6)$.

29. **(D)** $7x^3 + y^2 + 2x$. Remember that bases with unlike exponents cannot be combined.

30. **(C)** 24 square units. The bottom rectangle is $6 \times 2 = 12$, the middle one is $4 \times 2 = 8$, and the last one is $2 \times 2 = 4$, so the total area is $12 + 8 + 4 = 24$. Another way to look at this is as six squares, 2 units on a side, so the calculation is $6 \times 4 = 24$.

31. Less than. Amber makes $\$9 \times 20 = \$180$ per week. At that rate, she can have $\$1,800$ in $\dfrac{\$1,800}{\$180} = 10$ weeks, which is less than three months.

32. 3. An expression is undefined when the denominator of a fraction equals 0. For the first and third expressions, this occurs at $x = \pm 3$, for which the denominators are the difference of two squares. For $x = -3$, the middle expression is not undefined, since $(-6)^0 = 1$. But for $x = 3$, the middle expression is 0^0, which is undefined. Therefore, the answer is 3, but not -3. Note that anything to the 0 power equals 1, except that 0^0 is undefined. Also note that the numerators don't enter into determining when a fraction is undefined, only the denominators, which must not be 0.

33. 5. The diagonal divides the rectangle in half, so the area of the triangle formed is half the area of the rectangle, or $\left(\dfrac{1}{2} \times 4 \times 2\dfrac{1}{2}\right) = 2 \times 2\dfrac{1}{2} = 5$ square inches.

34. **(B)** 8. There are two ways to do this problem. For either approach, changing the $(4 - 2)$ in the parentheses to its equivalent, 2, is understood. The first is to evaluate the denominator (top) and divide the answer by 2, which will give $\dfrac{2^3 + 4^2 - 2^2 - (4 - 2)^2}{2} = \dfrac{8 + 16 - 4 - 4}{2} = \dfrac{16}{2} = 8$. The second method is to divide by 2 first (since it divides into every term), and then do the evaluation. So for the second approach, dividing 2 into each term yields $\dfrac{2^3 + 4^2 - 2^2 - (4 - 2)^2}{2}$. $2^2 + 2(4) - 2 - 2 = 8$. The tricky part of this solution is dividing 2 into 4^2. Since $4^2 = (2^2)(2^2)$, when we divide by 2, we divide it into only one of the factors, not both; 2 is divided into each term (separated by $+$ or $-$ signs), but only once per term, and 4^2 is one term, even though it contains two factors (which are multiplied). Therefore, the easier method is the first one.

35. 7/10. The probability that Jeff's name isn't selected is the product of the probabilities that someone else's name is selected in each of the three drawings. That would be $\dfrac{9}{10} \times \dfrac{8}{9} \times \dfrac{7}{8} = \dfrac{7}{10}$.

36. **(D)** $y = 3x - 2$. The equations in answer choices (A), (B), and (C) are all parallel to the given equation. They all have the same slope of 2. Therefore, they will never intersect, and there will be no solution. Only answer choice (D) will intersect with the given equation.

37. **(B)** $(2x + 6) = \dfrac{80}{2}$. The variable x represents the number of points scored by Billy. Felix's and Billy's scores together are $(x + 6) + x = 2x + 6$. The problem says that is half of the team's total of 80, so $2x + 6$ must equal $\dfrac{80}{2}$, or $(2x + 6) = \dfrac{80}{2}$.

38. Greater than. For an odd and then even selection with no replacement, the probability is $\dfrac{5}{10} \times \dfrac{5}{9} = \dfrac{5}{18}$.

    For an odd and then even selection with replacement, this becomes $\dfrac{5}{10} \times \dfrac{5}{10} = \dfrac{1}{4}$. Since $\dfrac{5}{18} = \dfrac{10}{36}$ and $\dfrac{1}{4} = \dfrac{9}{36}$, the answer is greater than.

39. **(D)** 48. This becomes a question of how many ways four people can sit on a bench because that one person is going to take an end seat. So the number of ways is $4 \times 3 \times 2 \times 1 = 24$. However, since there are two end seats, this has to be doubled to 48.

40. $-3, 4$ (in either order). Solve this most easily by substituting the given values when $x = 4$. Substituting -4 would give $(4 - 4) = 0$, so that choice is eliminated because anything times zero is zero. Substituting -3 would give $(4 - 3) = 1$ for one factor, and we just need the other factor to equal 8, so the other choice has to be 4, for an answer of $(x - 3)(x + 4) = (4 - 3)(4 + 4) = (1)(8) = 8$. No other combination of the given numbers will give 8 as an answer, since no others yield factors of 8 except -2, which needs the other choice to be 0 for $(4 - 2)(4 + 0) = 8$.

41. 2 units. The formula for the area of a rectangle is lw, where l is the length and w is the width. The first sentence says that $l = 4w$. So the area is

$$(4w)w = 16$$
$$4w^2 = 16$$
$$w^2 = 4$$
$$w = 2$$

42. **(B)** $F(-2) > F(-1) > F(3) > F(1)$. The values are $F(-2) = 17$, $F(-1) = 11$, $F(3) = 7$, $F(1) = 5$.

43a. 300. Jason's car gets $15 \times 20 = 300$ miles on a full tank.

43b. Three. Jason will have to stop for gas at 300, 600, and 900 miles into his trip, for a total of three times. The cost of gas, although given, isn't used for either answer.

44a. **(B)** (5, 5). Since this is a square, all side lengths are 4. Point O therefore is 4 units up and 4 units to the right of point M at (1, 1), so its coordinates are $(1 + 4, 1 + 4) = (5, 5)$.

44b. 1; -1. Slope is given by $\dfrac{\text{change in } y}{\text{change in } x}$. Since this is a square, all of the changes will be 4 or -4.

For the slope of MO, the changes in x and y are measured from M to O, which is $\dfrac{4}{4} = 1$. For the slope of NP, the changes in x and y are measured from N to P, which is $\dfrac{-4}{4} = -1$.

45. **(D)** It has no real roots. You might recognize right away that there is no real number whose square (which is always a positive number) added to 9 will equal zero. But since this is a quadratic, the discriminant ($b^2 - 4ac$) of the quadratic formula will also tell you the same thing, since it becomes $0^2 - 4(1)(9) = -36$, and there is a real root to a quadratic equation only if the discriminant is ≥ 0, or nonnegative.

Answer Sheet – Practice Test 2

1. Ⓐ Ⓑ Ⓒ Ⓓ

2. Ⓐ Ⓑ Ⓒ Ⓓ

3. _____

4. [_____]

5. Ⓐ Ⓑ Ⓒ Ⓓ

6. Ⓐ Ⓑ Ⓒ Ⓓ

7. [_____]

8. [_____]

9a. Ⓐ Ⓑ Ⓒ Ⓓ

9b. [_____]

9c. _____

10. Ⓐ Ⓑ Ⓒ Ⓓ

11. Ⓐ Ⓑ Ⓒ Ⓓ

12. [_____|_____]

13. Ⓐ Ⓑ Ⓒ Ⓓ

14. Ⓐ Ⓑ Ⓒ Ⓓ

15. [_____]

16. Ⓐ Ⓑ Ⓒ Ⓓ

17. Ⓐ Ⓑ Ⓒ Ⓓ

18. _____

19. Ⓐ Ⓑ Ⓒ Ⓓ

20. [_____]

21. Ⓐ Ⓑ Ⓒ Ⓓ

22. Ⓐ Ⓑ Ⓒ Ⓓ

23. _____

24a. _____

24b. [_____]

24c. Ⓐ Ⓑ Ⓒ Ⓓ

25. Ⓐ Ⓑ Ⓒ Ⓓ

26. Ⓐ Ⓑ Ⓒ Ⓓ

27. Ⓐ Ⓑ Ⓒ Ⓓ

28. Ⓐ Ⓑ Ⓒ Ⓓ

29. Ⓐ Ⓑ Ⓒ Ⓓ

30. Ⓐ Ⓑ Ⓒ Ⓓ

31. _____

32. _____

33. [_____]

34. Ⓐ Ⓑ Ⓒ Ⓓ

35. [_____]

36. Ⓐ Ⓑ Ⓒ Ⓓ

37. Ⓐ Ⓑ Ⓒ Ⓓ

38. _____

39. Ⓐ Ⓑ Ⓒ Ⓓ

40. _____

41. [_____]

42. Ⓐ Ⓑ Ⓒ Ⓓ

43a. [_____]

43b. [_____]

44a. Ⓐ Ⓑ Ⓒ Ⓓ

44b. [_____]

45. Ⓐ Ⓑ Ⓒ Ⓓ

Glossary

absolute value Positive distance of a number from 0 on the number line.

acute angle An angle whose measure is less than 90°.

acute triangle A triangle in which all the angles are acute angles, less than 90°.

algebraic expression Any combination of numbers and variables, such as $x^2 + 4x + 4$.

angle Two line segments or rays joined together at one of their endpoints. The measure of the angle tells how wide open it is.

arc The portion of the circumference of a circle between two points.

area The space enclosed by any figure.

average The sum of a given number of data points divided by the number of data points.

bar graph A graphic representation for comparing categorical data by means of rectangles with the same widths and with lengths proportional to the number of responses for each category.

base A number raised to a power. *Also,* one of the lines or surfaces of a geometric figure to which the altitude is constructed.

basic counting principle If there are *a* ways for one activity to happen, and *b* ways for a second activity to happen, then there are $a \times b$ ways for both *a* and *b* to happen.

binomials Mathematical expressions with two terms, such as $x + y$ or $x - 2$.

boxplot A box drawn above a number line that provides ready information on the center

(median) and variation (quadrants, range) in a data set.

cancellation A method to reduce fractions to more manageable forms by canceling factors in the numerator and denominator.

capacity *See* volume.

Cartesian coordinate Location of a point on a plane; the distances of the point from each of two intersecting straight-line axes of reference.

categorical data Data that relate to categories, such as "yes/no," or "black, white, red."

center (of a circle) The point at which the distance from every point on the circumference of the circle is the same.

center (of a sphere) The point at which the distance from every point on the surface of the sphere is the same.

central angle An angle with the center of a circle as its vertex and two radii as its sides.

central tendency Data that tell us how the center of a group of data tends to look.

circle graph *See* pie chart.

circle All the points at a fixed distance from a certain point.

circumference The perimeter of a circle; all the points at a fixed distance from the center.

closed figure A geometric figure in which all the corners are connected.

coefficients The numbers in front of the variables; for example the 3 in $3x^2$.

combinations How many ways items can be arranged when order doesn't matter.

common denominator A common multiple of the denominators of a number of fractions. A *lowest* common denominator is the smallest of these multiples.

commutative property The sum (or product) of a and b is the same as the sum (or product) of b and a.

complementary angles Two angles whose measures add up to $90°$.

composite figures Geometric figures made up of two or more shapes.

composite function Functions made up of two or more different functions.

cone A three-dimensional figure with a circle at one end and a point at the other.

correlation An interdependence between mathematical variables.

cost The amount a seller pays to a wholesaler.

cross-multiplication Multiplication in a proportion (presented as two equal fractions) in the form of a cross ($\times$), making the two products equal.

cube A three-dimensional figure that has six square faces. *Also,* the value when a number or variable is a factor three times in a multiplication. For example the cube of a is $a^3 = a \times a \times a$.

cube root The factor that equals a variable or a number when multiplied three times by itself. The cube root of x^3 is x, and the cube root of 8 is 2.

cubed Another way to say "to the third power."

cylinder A three-dimensional figure that is in the shape of a can, with identical circles on each end.

data Information that can have many forms: numbers, charts, tables, names, and political parties, just to name a few.

denominator The denominator, or bottom number, in a fraction.

diameter Twice the radius of a circle or a sphere.

discount A deduction (usually a percentage) from the price of an item.

discriminant The expression under the radical in the quadratic equation, $b^2 - 4ac$.

domain The value of x on an xy coordinate system.

dot plot *See* scatter plot.

edges The lines where two faces meet on a three-dimensional figure.

endpoint The end of a line segment or of a drawn graph.

equation The algebraic statement of the equality of two expressions.

equiangular triangle A triangle in which all three angles are equal.

equilateral triangle A triangle in which all three sides are equal.

exponent A symbol written above and to the right of a symbol, expression, or quantity to indicate how many times it is to be taken as a factor.

exponential function A function that typically has a slow growth and then an accelerated growth, or a slow decay and then an accelerated decay.

faces The flat surfaces that meet on a three-dimensional figure.

factorial Multiplication of factors from a given number down to 1.

F-O-I-L method Mnemonic for multiplying two binomials: first, outside, inside, last.

fraction How many parts of a whole thing, expressed as a numerator (dividend) over a denominator (divisor).

frequency curve A plot of data versus frequency.

function A special relationship between input values and output values such that for each input value, there can be only one output value.

height The perpendicular distance from the base to the opposite side or opposite angle of a geometric figure.

histogram A way of showing the distribution of data. The horizontal axis shows the data in equal-size groups and the vertical axis shows the frequency of each group.

hypotenuse The side opposite the 90° angle in a right triangle; it is the longest side.

improper fractions A fraction in which the numerator is larger than the denominator.

index A smaller (in size) number at the left of a radical sign that indicates the root.

inequalities A statement between two quantities signifying that one is less than, greater than, or not equal to the other.

integer Any of the natural numbers (1, 2, 3, 4, 5, . . .), the negatives of these numbers, and 0.

intersection The set of points common to two geometric configurations, such as the point at which two lines cross each other.

inverse An opposite operation. The inverse of addition is subtraction (and vice versa), and the inverse of multiplication is division (and vice versa).

inverting a fraction Reversing the numerator and denominator.

isosceles triangle A triangle with two equal side lengths.

like terms Terms that have exactly the same variables to the same powers.

line A one-dimensional figure made up of points that that go on to infinity (∞) in both directions.

line segment A part of a line with a definite length.

lowest terms A fraction that cannot be reduced any further.

markdown *See* discount.

markup The amount added to the cost to cover expenses and profit.

mean The sum of a given number of data points divided by the number of data points.

median The value for which half the data are above and half are below.

midpoint The point that is exactly halfway between the endpoints of a line segment.

mixed number A number consisting of a whole number and a fraction, such as $2\frac{4}{5}$.

mnemonic A technique for aiding memory, such as the made-up word PEMDAS to remember the order of operations:

parentheses, exponents, multiplication, division, addition, and subtraction.

mode The most common value in a data set.

monomials Mathematical expressions with one term.

mutually exclusive A situation in which there is no chance that one item fits the criteria of another item.

negative exponent An indication of reciprocal.

numerator The top number in a fraction; the dividend.

obtuse angle An angle whose measure is greater than 90° and less than 180°.

obtuse triangle A triangle in which one of the angles is an obtuse angle.

order of operations Parentheses, exponents, multiplication, division, addition, subtraction.

ordered pair (*x, y*) Any point on the Cartesian coordinate plane in which the *x* indicates the horizontal value and the *y* indicates the vertical value.

ordinal data Data that relate to rank, such as "first, second, third."

origin The point where the axes on the Cartesian coordinate system meet.

outlier A data point that is clearly outside of the range of the other data points.

parabola The distinct shape of the graph of a quadratic equation.

parallel lines Lines that are an equal distance apart and go on forever, never meeting or crossing.

parallelogram A quadrilateral (four-sided figure) with two pair of parallel sides.

percentage A part of a whole expressed in hundredths.

perimeter The distance around a figure, or the sum of the lengths of all of its sides.

periodicity A repeat of a graph at regular intervals.

permutation How many ways items can be arranged when order matters.

perpendicular lines Lines that meet, or intersect, at a right angle (90°).

pi (π) The ratio of the circumference of a circle to its diameter.

pie chart A circle divided into wedges that are proportional to the percentages in a data set.

point-slope form An expression of the equation of a line, $(y_2 - y_1) = m(x_2 - x_1)$, in which any two points on the line are (x_1, y_1) and (x_2, y_2) and the slope is m.

polygon A closed figure consisting of straight lines joined end to end.

polynomials Mathematical expressions with two or more terms.

population The items that are of interest in statistics, not necessarily people.

power The number of repeated factors.

price The sum of cost plus markup.

prime number Number that doesn't have factors other than 1 and itself.

probability A value between 0 and 1 that indicates the chance that a given event will occur.

proportion The way to express that two ratios are equal.

Pythagorean Theorem The square of the hypotenuse is equal to the sum of the squares of the other two sides, or $c^2 = a^2 + b^2$, where c is the length of the hypotenuse of a right triangle, and a and b are the lengths of the two legs.

Pythagorean triples The sides of right triangles that are whole numbers, such as 3, 4, 5.

quadrants The four parts of the Cartesian coordinate system delineated by the x- and y-axes.

quadratic An expression in which the highest power of the unknown is 2.

quadratic equation An equation in which the unknown is squared and there is no higher power of the unknown.

quadratic formula Formula for finding the roots of a quadratic equation.

quadrilateral A four-sided closed figure.

quantitative data Data that are numerical.

radical sign The symbol for root, $\sqrt{}$.

radicand The number that goes under the radical sign for which the root is taken.

radius The fixed distance from the center to the circumference of a circle or surface of a sphere.

random sample Every member in the population has an equal chance of being picked for the sample.

range The value of y on an xy coordinate system. *Also,* the highest value minus the lowest value in a data set.

ratio A way of comparing two numbers or expressing the relation between two quantities.

rational number Any number that can be written as a fraction.

ray A line that goes off to ∞ in only one direction. The other end has a definite point, or endpoint.

real number Any number that doesn't have $\sqrt{-1}$ as a factor.

regular polygon A polygon in which all the sides are equal.

reciprocal The number formed by the fraction that has 1 as the numerator and the given number as the denominator.

rectangle A parallelogram with four equal angles.

rectangular prism A three-dimensional figure with perpendicular edges and faces.

regular solids A three-dimensional figure in which all sides and the base are identical regular polygons.

relative maximum The highest point on a graph relative to the points on either side of it.

relative minimum The lowest point on a graph relative to the points on either side of it.

rhombus A quadrilateral with all the properties of a parallelogram plus equal sides.

right angle The angle formed by perpendicular lines; an angle that contains 90°.

right prism *See* rectangular prism.

right regular pyramid A three-dimensional figure in which one base is a regular shape (all the sides are equal), the other end is a point, and all of the other faces are identical triangles.

right triangle A triangle in which one of the angles is a right angle, or 90°.

root Solution to a quadratic equation. *Also,* the opposite of power.

rounding Expressing a number by dropping decimals beyond a stated number of places or by substituting zeros for final integers.

sale price Price minus discount.

scale factor Factor by which all dimensions of a geometric figure are multiplied.

scalene triangle A triangle in which the lengths of all of the sides are different.

scatter plot A plot of one variable against another to see whether there is an association between them.

scientific notation A method that uses powers of 10 to write numbers that are either too big or too small to be conveniently written as decimals.

semicircle A half circle.

simplifying a fraction Reducing a fraction to its lowest terms by eliminating common factors from the numerator and denominator.

simplifying roots A method that uses factoring of the radicand into a product with a perfect square.

simultaneous equations Equations with more than one unknown that have one point at which all the equations are true.

skewed data Data with extreme scores at one end of the ordered data.

slant height The measure from the point of a three-dimensional figure (e.g., cone, pyramid) to the middle of the side of the base.

slope of a line $m = \dfrac{y_2 - y_1}{x_2 - x_1} = \dfrac{\text{change in } y}{\text{change in } x}$, where (x_1, y_1) and (x_2, y_2) are any two points on the line.

slope-intercept form An expression of the equation of a line as $y = mx + b$, where m is the slope of the line and b is the y-intercept.

sphere A three-dimensional figure in which every point on the surface of the sphere is the same distance from a point called the center of the sphere.

square A rectangle with four equal sides; can also be thought of as a rhombus with four equal angles.

square root A factor of a number that when squared gives the number; for example, the square roots of 4 are +2 or −2.

squared To the second power.

statistics A way of presenting information about a set of data.

supplementary angles Two angles whose measures add up to 180°.

surface area The total area of all the faces of a three-dimensional figure, even those that cannot be seen.

systems of equations *See* simultaneous equations.

term Each part of an expression separated by a + or − sign.

three-dimensional figures Geometric figures that have depth in addition to length and width.

trapezoid A quadrilateral with only two sides that are parallel.

triangle Three line segments that form a closed figure.

trinomials Mathematical expressions with three terms.

unit rate How many units of one variable correspond to one unit of another variable.

variables The letters used for the unknown quantities.

vertex The highest point of a parabola. *Also,* the point of an angle.

vertical angles The angles across from each other when two lines cross.

volume How much a three-dimensional figure can hold.

whole numbers Any of the natural numbers (1, 2, 3, 4, 5, . . .), the negatives of these numbers, and 0.

x-**axis** The horizontal axis in Cartesian coordinates.

y-**axis** The vertical axis in Cartesian coordinates.

Index

A

absolute maximum, 154
absolute value, 35–38
acute angle, 168
acute triangle, 173
addition, 30–39
    associative property, 39
    carrying, 33–34
    commutative property, 39
algebraic expressions, 109–114
    addition, 110
    division, 110–111
    evaluating, 114
    exponent rules, 111–112
    multiplication, 110–111
    subtraction, 110
angles, 167–170
arc, 189
area, 167
    of a circle, 190
    of a parallelogram, 183
    of a rectangle, 184
    of a rhombus, 183
    of a square, 185
    of a trapezoid, 182
    of a triangle, 179
associative property, 39, 45
average, 225

B

bar graph, 236–237
base
    with exponents, 87, 89–90, 111
    of a pyramid, 200
    of a triangle, 172, 178
basic counting principle, 215–216
binomials, 141
boxplots, 235

C

calculator instructions 11–25
cancellation, 63–64, 67
capacity. *See* volume
Cartesian coordinates, 114–115
center
    of a circle, 188
    of a sphere, 204
central angle, 189–190
central tendency, 225–227
circle graph. *See* pie chart
circle, 188–192
circumference, 190
closed figures, defined, 167
coefficients, 110–111
combinations, 217–220
common denominator, 69
    lowest, 69–70
commutative property, 39, 44, 48, 63
complementary angles, 169
composite figures
    three-dimensional, 206–207
    two-dimensional, 193
composite function, 154
cone, 202–203
correlation, 237–238
cost, 76–78
cross-multiplication, 112
cube root, 91
cube, 198
cubed, 87
cylinder, 199

D

data, 224–230
    categorical, 224
    central tendency, 225–227
    ordinal, 224
    quantitative, 225
    skewed, 227–228
    of, 230
    statistical displays, 232–238
decimal system, 30–31
decimals, 53, 55–58
    addition, 55
    changing to fractions, 75
    changing to percentages, 75–76
    division, 56–57
    multiplication, 56
    subtraction, 55
denominator, 59–70, 73–75
diameter, 188
difference of squares, 146
discount, 76–78
discriminant, 147
distance-time graphs, 119–121
distributive property, 42, 45
division, 45–48
domain, 150
dot plot. *See* scatter plot

E

edges, 194–201
endpoint
    of graph, 154
    of line segment, 168
equations, 103–109
equiangular triangle, 172, 179
equilateral triangle, 172, 179
exponential function, 153–154
exponents, 88–90
    fractional, 92
    negative, 88–89, 111
    rules for, 111–112
    See also powers

NOTES

NOTES

NOTES

NOTES

NOTES

NOTES

NOTES

NOTES

NOTES

NOTES